An NPR Best Book of 2015

A *Washington Post* Notable Nonfiction Book of 2015

An Orwell Prize Finalist

An *Independent* Book of the Year

A *New Statesman* Book of the Year

PRAISE FOR *THE DARK NET*

"Bracing . . . An excellent book . . . Bartlett combines an insider's expertise with a neophyte's tale of discovery. Rather than measure the pros and cons of the Web, he maps its frontiers without judgment. The result is a lucid inquiry into the relationship between technology and freedom that's also a captivating beach book."

—*The Washington Post*

"A hell of an achievement . . . Buy it and read it."

—*The Times* (London)

"Bartlett anatomises the usual bogeymen and demonstrates that they're real. *The Dark Net* is, for anyone engaged with the web and the effects it is having on our culture, necessary reading . . . A flashlight in a dark, dark cellar." —*The Spectator*

"Fascinating . . . A confident and well-informed guide . . . By meeting the people behind the online activity, Bartlett humanizes it."

—*New Scientist*

"Reveals a hidden, seedy world where people lurk behind pseudonyms and dupe others into revealing their bodies on camera to be used against them in public shaming. If you're shocked to discover that last year approximately 20 per cent of drug users bought their stash online, you'll find this fascinating. Bartlett is an able guide on a journey through the margins of the web."

—*The Independent*

"It is Bartlett's plentiful and fascinating interviews with the denizens of the dark net that make his book so compelling . . . Quite worrying, a bit disgusting, highly voyeuristic, and occasionally very funny: this is the nature of both the dark net and *The Dark Net*."

—*Barnes & Noble Review*

"Bartlett . . . began his breezy and humane book expecting to find more clear-cut cases of people and behaviors to condemn and proscribe. But what he ended up finding was just human beings: dedicated, troubled, incendiary, funny, craving to be heard and understood. Welcome to the Dark Net. Welcome to the human race."

—*Reason*

"*The Dark Net* offers smart, provoking reportage from the crooked crannies of digital culture, married to a quietly impressive analysis of how technology is amplifying both the best and the worst of us. Required reading for anyone looking to escape media hysteria and get to grips with the 21st century's most compelling, discomforting complexities."

—Tom Chatfield, author of *Fun Inc.: Why Gaming Will Dominate the Twenty-First Century*

"Fascinating . . . A provocative journey through the deep web's history, its varied guiding philosophies, and the bizarre, iconoclastic, often criminal behaviors it conceals and energizes."

—*The Brooklyn Rail*

"A welcome deep dive into the anonymous Internet."

—*Flavorwire*

THE DARK NET

INSIDE THE DIGITAL UNDERWORLD

THE DARK NET

JAMIE BARTLETT

BROOKLYN • LONDON

THE DARK NET

First published in Great Britain in 2014 by William Heinemann, an imprint of Random House

First published in the United States in 2015 by Melville House

First Melville House paperback printing: May 2016

Melville House Publishing
46 John Street
Brooklyn, NY 11201

and

8 Blackstock Mews
Islington
London N4 2BT

mhpbooks.com facebook.com/mhpbooks @melvillehouse

Paperback ISBN: 978-1-61219-521-6

Printed in the United States of America

10 9 8 7 6 5 4

Design by Adly Elewa

A catalog record for this book is available from the Library of Congress

For Huey, Max, Sonny, and Thomas, who were born while I was writing this book. When they are old enough, I hope they will read it and wonder what on earth all the fuss was about, and laugh at their uncle's hopeless predictions.

CONTENTS

Author's Note xi

Introduction: Liberty or Death 3

1. Unmasking the Trolls 14
2. The Lone Wolf 48
3. Into Galt's Gulch 73
4. Three Clicks 110
5. On the Road 135
6. Lights, Web-camera, Action 166
7. The Werther Effect 193

Conclusion: Zoltan vs. Zerzan 219

Acknowledgments 241

Notes 245

Further Reading 301

AUTHOR'S NOTE

The Dark Net is an examination of what are, in many cases, extremely sensitive and contentious subjects. My primary aim was to shine a light on a world that is frequently discussed, but rarely explored—often for good reason. Throughout I have endeavored to set my own views aside and write as objective and as lucid an account of what I experienced as possible. Some readers may question the wisdom of writing about this subject at all, and express concern at the information *The Dark Net* reveals. Although my intention was never to provide a guide to illegal or immoral activity online, this book does contain material that some will find shocking and offensive.

As a researcher I felt a duty to respect the privacy of the people I encountered. Where necessary, I have altered names, online pseudonyms, and identifying details, and, in one chapter, created a composite character based on several individuals. For the reader's ease, I have also corrected many (but not all) spelling mistakes in quoted material.

I have tried to balance the rights of individuals with the social benefit that I believe comes from describing them and the worlds they inhabit. It is not a foolproof method; rather a series of judgments. Any errors, omissions, and mistakes are mine alone, and I hope those included in this book will accept my apologies in advance for any distress or discomfort caused.

Online life moves quickly. By the time you read *The Dark Net*, certain parts of these stories will have changed, websites will have closed down, subcultures will have evolved, new laws will have been enacted. But its core theme—what humans do under the conditions of real or perceived anonymity online—will certainly have not.

JAMIE BARTLETT

March 2015

THE DARK NET

INTRODUCTION

LIBERTY OR DEATH

I have heard rumors about this website, but I still cannot quite believe that it exists. I am looking at what I think is a hit list. There are photographs of people I recognize—prominent politicians, mostly—and, next to each, an amount of money. The site's creator, who uses the pseudonym Kuwabatake Sanjuro, thinks that if you could pay to have someone murdered with no chance—I mean absolutely zero chance—of being caught, you would. That's one of the reasons why he has created the Assassination Market. There are four simple instructions listed on its front page:

>Add a name to the list

>Add money to the pot in the person's name

>Predict when that person will die

>Correct predictions get the pot

The Assassination Market can't be found with a Google search. It sits on a hidden, encrypted part of the internet that, until recently, could only be accessed with a browser called The Onion Router, or Tor. Tor began life as a U.S. Naval Research Laboratory project, but today exists as a not-for-profit organization, partly funded by the U.S. government and various civil liberties groups, allowing millions of people around the world to browse the internet anonymously and securely.* To put it simply, Tor works by repeatedly encrypting computer activity and routing it via several network nodes, or "onion routers," in so doing concealing the origin, destination, and content of the activity. Users of Tor are untraceable, as are the websites, forums, and blogs that exist as Tor Hidden Services, which use the same traffic encryption system to cloak their location.

The Assassination Market may be hosted on an unfamiliar part of the net, but it's easy enough to find, if you know how to look. All that's required is simple (and free) Tor software. Then sign up, follow the instructions, and wait. It is impossible to know the number of people who are doing exactly that, but at the time of writing, if I correctly predict the date of the death of Ben Bernanke, the former chairman of the Federal Reserve, I'd receive approximately $56,000.

It may seem like a fairly pointless bet. It's very difficult to guess when someone is going to die. That's why the Assassination Market has a fifth instruction:

>Making your prediction come true is entirely optional

THE DARK NET

The Assassination Market is a radical example of what people do online when under the cover of real or perceived anonymity.

* In 2010, Tor was awarded the Free Software Foundation's Award for Projects of Social Benefit, in part for the service it provides for whistleblowers, human-rights campaigners, and activists in dissident movements.

Beyond the more familiar world of Google, Hotmail, and Amazon lies another side to the internet: the dark net.

For some, the dark net refers to the encrypted world of Tor Hidden Services, where users cannot be traced, and cannot be identified. For others, it is those sites not indexed by conventional search engines: an unknowable realm of password-protected pages, unlinked websites, and hidden content accessible only to those in the know, sometimes referred to as the "deep web." It has also become a catchall term for the myriad shocking, disturbing, and controversial corners of the net—the realm of imagined criminals and lurking predators.

The dark net, for me, describes an idea more than a particular place: internet underworlds set apart yet connected to the internet we inhabit, worlds of freedom and anonymity, where users say and do what they like, often uncensored, unregulated, and outside of society's norms. It is dark because we rarely see these parts of digital life, save the occasional flash of a hysterical news report or shocking statistic. This is not a book about Tor, since the net is full of obscure corners, of secret back alleys on parts of the internet you likely already know: social media sites, normal websites, forums, chat rooms. I focus instead on those digital cultures and communities that appear, to those that aren't part of them, dark, insidious, and beyond society's gaze—wherever I found them.

This dark net is rarely out of the news—with stories of young people sharing homemade pornography, of cyberbullies and trolls tormenting strangers, of hackers stealing and leaking personal photos, of political or religious extremists peddling propaganda, of illegal goods, drugs, and confidential documents only a click or two away appearing in headlines almost daily—but it is still a world that is, for the most part, unexplored and little understood.

In reality, few people have ventured into the darker recesses of the net to study these sites in any detail.

I started researching radical social and political movements in 2007, when I spent two and a half years following Islamist extremists around Europe and North America, trying to piece together a fragmented and largely disjointed real-world network of young men who sympathized with al-Qaeda ideology. By the time I'd finished my work in 2010, the world seemed to be different. Every new social or political phenomenon I encountered—from conspiracy theorists to far-right activists to drug cultures—was increasingly located and active online. I would frequently interview the same person twice—once online and then again in real life—and feel as if I was speaking to two different people. I was finding parallel worlds with different rules, different patterns of behavior, different protagonists. Every time I thought I'd reached the bottom of one online culture, I discovered other connected, secretive realms still unexplored. Some required a level of technical know-how to access, some were extremely easy to find. Although an increasingly important part of many people's lives and identities, these online spaces are mostly invisible: out of reach and out of view. So I went in search of them.

My journey took me to new places online and offline. I became the moderator of an infamous trolling group and spent weeks in forums dedicated to cutting, starving, or killing yourself. I explored the labyrinthine world of Tor Hidden Services in search of drugs, and to study child pornography networks. I witnessed online wars between neo-Nazis and antifascists on popular social media sites, and signed up to the latest porn channels to examine current trends in homemade erotica. I visited a Barcelona squat with anarchist Bitcoin programmers, run-down working men's clubs to speak to extreme nationalists, and a messy bedroom to observe three girls make a small fortune performing sexually explicit acts on camera to thousands of viewers. By exploring and comparing these worlds, I also hoped to answer a difficult question: do the features of anonymity and connectivity free the darker sides of our nature? And if so, how?

The Dark Net is not an effort to weigh up the pros and cons of the internet. The same anonymity that allows the Assassination Market to operate also keeps whistleblowers, human-rights campaigners, and activists alive. For every destructive subculture I examined there are just as many that are positive, helpful, and constructive.

Because the internet has become so interwoven into the fabric of our lives, it presents a challenge to our existing notions of anonymity, privacy, freedom, and censorship—throwing up new challenges not yet resolved: should we have the right to complete anonymity online? Are our "digital" identities distinct from our "real" ones—and what does that mean? Are we prone to behave in particular ways when we sit behind a screen? What are the limits of free expression in a world where every idea is a click away? Particularly since the revelations of the former National Security Agency contractor Edward Snowden, these questions dominate debates and discussion about the role of internet privacy and freedom in an increasingly digital world. I don't propose any easy answers or solutions. I'm not sure that there are any. This book is not a polemic—more modestly, it is a series of portraits about how these issues play out at the fringes. I leave it entirely to you to decide what you think it means.

CONNECTED

The net as we know it started life in the late 1960s, as a small scientific project funded and run by the Advanced Research Projects Agency (ARPA), a development arm of the U.S. military. The Pentagon hoped to create an "Arpanet" of linked computers to help top American academics share data sets and valuable computer space. In 1969, the first networked connection was made between two computers in California. It was a network that slowly grew.

In July 1973, Peter Kirstein, a young professor of computer science at University College London, connected the UK to the Arpanet via the Atlantic seabed phone cables, a job that made Kirstein the first person in the UK online. "I had absolutely no idea what it would become!" Kirstein tells me. "None of us did. We were scientists and academics focused on trying to build and maintain a system which allowed data to be shared quickly and easily." The Arpanet, and its successor, the internet, was built on principles that would allow these academics to work effectively together: a network that was open, decentralized, accessible, and censorship-free. These ideas would come to define what the internet stood for: an unlimited world of people, information, and ideas.

The invention of Bulletin Board Systems (BBS) in 1978, and Usenet in 1979–80, introduced a new generation to life online. Unlike the cloistered Arpanet, Usenet and BBS, the forerunners of the chat room and forum, were available to anyone with a modem and a home computer. Although small, slow, and primitive by today's standards, they were attracting thousands of people intrigued by a new virtual world. By the mid-nineties and the emergence of Tim Berners-Lee's World Wide Web, the internet was fully transformed: from a niche underground haunt frequented by computer hobbyists and academics, to a popular hangout accessed by millions of excited neophytes.*

According to John Naughton, Professor of the Public Understanding of Technology at the Open University, cyberspace at this time was more than just a network of computers. Users saw it as "a new kind of place," with its own culture, its own identity, and its own rules. The arrival of millions of "ordinary" people online stimulated fears and hopes about what this new form of communication might do to us. Many techno-optimists, such as the cheerleaders for the networked revolution *Wired* and *Mondo 2000* magazines, believed cyberspace would herald a new dawn of learning and understanding, even the end of the national state. The best statement of this view was the American essayist and prominent cyberlibertarian John Perry Barlow's 1996 "Declaration of the Independence of Cyberspace," which announced to the real world that "your legal concepts of property, expression, identity, movement, and context do not apply to us . . . our identities have no bodies, so, unlike you, we cannot obtain order by physical coercion." Barlow believed that the lack of censorship and the anonymity that the net seemed to offer would foster a freer, more open society, because people could cast off the tyranny of their fixed real-world identities and create themselves anew. (*The New Yorker* put it more succinctly: "On the Internet, nobody knows you're a dog.") Leading psychologists of the day, such as MIT professor Sherry Turkle in her influential 1995 study of internet identity, *Life on the Screen*, offered a cautious welcome to

* September 1993, the month America Online started to offer its subscribers access to Usenet, is etched into internet folklore as "the eternal September," when newcomers logged on to the internet en masse.

the way that online life could allow people to work through the different elements of their identity.

But others worried what might happen if no one knows you're a dog. Parents panicked about children infected with "modem fever." Soon after Turkle's study, another psychologist, John Suler, was studying the behavior of participants in early chat rooms. He found that participants tended to be more aggressive and angry online than offline. He suggested this was because, when protected by a screen, people feel that real-world social restrictions, responsibilities, and norms don't apply. Whether actual or perceived, anonymity, thought Suler, would allow you to explore your identity, but it might also allow you to act without fear of being held accountable (in 2001 he would call this "The Online Disinhibition Effect"). It's true that from the outset, many BBS and Usenet subscribers were treating cyberspace as a realm for all sorts of bizarre, creative, offensive, and illegal behavior. In Usenet's "Alternative" hierarchy, anyone could set up a discussion group about anything they wanted. The first group was alt.gourmand, a forum for recipes. This was swiftly followed by alt.sex, alt.drugs and alt.rock-n-roll. "*Alt.*," as it came to be known, immediately became the most popular part of Usenet by far. Alongside purposeful and serious groups for literature, computing, or science, Usenet and BBS contained many more dedicated to cyberbullying, hacking, and pornography.

GIVE ME LIBERTY OR GIVE ME DEATH

It was in this heady atmosphere that the radical libertarian Jim Bell first took the promise of online anonymity to a terrifying conclusion. In late 1992, a group of radical libertarians from California called the "cypherpunks" set up an email list to propose and discuss how cyberspace could be used to guarantee personal liberty, privacy, and anonymity. Bell, a contributor to the list, believed that if citizens could use the internet to send secret encrypted messages and trade using untraceable currencies, it would be possible to create a functioning market for almost anything. In 1995, he set out his ideas in an essay called "Assassination Politics," which he posted to the email list. It made even the staunchly libertarian cypherpunks wince.

Bell proposed that an organization be set up that would ask citizens to make anonymous digital cash donations to the prize pool of a public figure. The organization would award the prize to whoever correctly predicted that person's death. This, argued Bell, wasn't illegal, it was just a type of gambling. But here's the ruse: if enough people were sufficiently angry with a particular individual—each anonymously contributing just a few dollars—the prize pool would become so large that someone would be incentivized to make a prediction and then fulfil it themselves in order to take the pot. This is where encrypted messages and untraceable payment systems come in. A crowd-sourced—and untraceable—murder would unfold as follows. First, the would-be assassin sends his prediction in an encrypted message that can be opened only by a digital code known to the person who sent it. He then makes the kill and sends the organization that code, which would unlock his (correct) prediction. Once verified by the organization, presumably by watching the news, the prize money—in the form of a digital currency donated to the pot—would be publicly posted online as an encrypted file. Again, that file can be unlocked only by a "key" generated by whoever made the prediction. Without anyone knowing the identity of anyone else, the organization would be able to verify the prediction and award the prize to the person who made it.

The best bit, thought Bell, was that internet-enabled anonymity safeguarded all parties, except perhaps the killer (and his or her victim). Even if the police discovered who'd been contributing to the cash prizes of people on the list, the donors could truthfully respond that they had never directly asked for anyone to be killed. The organization that ran the market couldn't help either, because they wouldn't know who had donated, who had made predictions or who had unlocked the cash file. But Bell's idea was about more than getting away with murder. He believed that this system would exert a populist pressure on elected representatives to be good. The worse the offender—the more he or she outraged his or her citizens—the more likely they were to accumulate a large pool, and incentivize potential assassins. (Bell believed Stalin, Hitler, and Mussolini would all have been killed had such a market existed at the time.) Ideally, no one would need to be killed. Bell hoped the very existence of this market would mean no one would dare throw their hat into the ring at all. "Perfect anonymity, perfect secrecy, and perfect security," he wrote, ". . . combined with the ease and security with which these contributions could be collected, would make being an abusive government employee an extremely risky proposition. Chances are good that nobody above the level of county commissioner would even risk staying in office."

In 1995, when Bell wrote "Assassination Politics," this was all hypothetical. Although Bell believed his market would ultimately lead to the collapse of every government in the world, reality hadn't caught up with his imagination. Nearly two decades later, with the creation of digital currencies like Bitcoin, anonymous browsers like Tor and trustworthy encryption systems, it had, and Bell's vision was realized. "Killing is in most cases wrong, yes," Sanjuro wrote when he launched the Assassination Market in the summer of 2013:

> However, this is an inevitable direction in the technological evolution . . . When someone uses the law against you and/or infringes upon your rights to life, liberty, property, trade or the pursuit of happiness, you may now, in a safe manner from the comfort of your living room, lower their life-expectancy in return.

There are, today, at least half a dozen names on the Assassination Market. Although it is frightening, no one, as far as I can tell, has been assassinated. Its significance lies not in its effectiveness, but in its existence. It is typical of the sort of creativity and innovation that characterizes the dark net: a place without limits, a place to push boundaries, a place to express ideas without censorship, a place to sate our curiosities and desires, whatever they may be. All dangerous, magnificent, and uniquely human qualities.

1
UNMASKING THE TROLLS

At the top of the tree of life there isn't love: there is lulz.

—Anonymous

A LIFE RUIN

"Hi /b/!" read the small placard that Sarah held to her semi-naked body. "7 August 2013, 9:35 p.m."

It was an announcement to the hundreds—thousands, perhaps—of anonymous users logged on to the infamous "/b/" board on the image-sharing website 4chan that she was ready to "cam." Appreciative viewers began posting various sexually explicit requests, which Sarah performed, photographed, and uploaded.

On 4chan, there are boards dedicated to a variety of subjects, including manga, DIY, cooking, politics, and literature. But the majority of the twenty million people from all over the world who visit the site each month head for /b/, otherwise known as the "random" board. Sarah's photographs were only part of one of many bizarre, offensive, or sexually graphic image "threads" constantly running on /b/. Here, there is little to no moderation, and almost everyone posts anonymously. There is, however, a set of loose guidelines: the *47 Rules of the Internet*, created by /b/users, or "/b/tards," themselves, including:

Rule 1: Do not talk about /b/
Rule 2: Do NOT talk about /b/
Rule 8: There are no real rules about posting
Rule 20: Nothing is to be taken seriously
Rule 31: Tits or G[et] T[he] F[uck] O[ut]—the choice is yours
Rule 36: There is always more fucked-up shit than what you just saw
Rule 38: No real limits of any kind apply here—not even the sky
Rule 42: Nothing is sacred

The anonymous and uncensored world of /b/ generates an enormous amount of inventive, funny, and offensive content, as users vie for popularity, and notoriety. Did you ever click on a YouTube link and unexpectedly open Rick Astley's 1987 smash hit "Never Gonna Give You Up"?* That was /b/. Or receive funny photographs of cats with misspelled captions? Also /b/. The hacktivist group Anonymous? /b/ again. All those naked photos of celebrities that were posted online in the summer of 2014? /b/.

But anonymity has its downside. Female users are a novelty here, and are routinely ignored or insulted, that is, unless they post photographs of themselves, or play "camgirl," which is always

* If you did, you were one of over twenty million people who were "rickrolled" that year.

a simple and effective way to capture the attention of the /b/tards. 4chan has a dedicated board for camming, called "/soc/," where users are expected to treat camgirls nicely. Every day, dozens of camgirls appear there and perform. But occasionally one foolishly strays into /b/.

Approximately twenty minutes after the first photograph was posted, one user requested that Sarah take a naked photograph of herself with her first name written somewhere on her body. Soon afterwards, another user asked for a naked photograph of her posing with any medication she was taking. She duly performed both tasks. This was a mistake.

> **Anonymous said:** shit, I hope no one doxxes her. She actually delivered. She seems like a kind girl.
>
> **Anonymous replied:** dude get a grip she gave her first name, her physician's full name, and even the dormitory area she lives in she wants to be found.
>
> **Anonymous replied:** She is new. Any girl who makes signs or writes names on her body is clearly new to camwhoring, so they really don't know what they're getting themselves into.

Sarah had inadvertently provided enough personal information to allow users to "dox" her—to trace her identity. Other /b/tards were alerted and quickly joined the thread—on 4chan, doxing a camwhore is seen as a rare treat—and before long, users had located Sarah on her university's searchable directory (based somewhere in the United States) and revealed her full name, address, and telephone number. Next, they tracked down her Facebook and Twitter accounts. Sarah was still at her computer, watching helplessly.

> **Anonymous said:** STOP. Seriously. Fucking fat losers
>
> > **Anonymous replied:** good to see you're still in the thread sarah. You're welcome btw.
> >
> > > **Anonymous replied:** heyyy . . . sarah . . . can I add you on facebook? Just kidding delete that shit before your nudes get sent to your friends
>
> **Anonymous said:** She literally just made her fucking twitter private while I was browsing her pics. Fucking cunt.
>
> > **Anonymous replied:** It's K if she does delete it. I'm making notes on the people on her friends list and their relation with her. Will start sending the nudes soon.
> >
> > > **Anonymous replied:** LOL she deleted her Facebook. Doubt she can delete her relatives though.
> > >
> > > > **Anonymous replied:** Eh, just save her name. Eventually once all this settles she will reactivate it and she will have her jimmies rustled once more. She will now never know peace from this rustling. And she's going to have one embarassing fucking time with her family.

Anonymous said: You fucking nerdbutts got her Facebook? You guys are fucking unbelievable. A girl actually delivers on this shit site, and you fuckers dox her. Fucking /b/, man.

> **Anonymous replied:** get the fuck out you piece of shit moralfag trash
>
> > **Anonymous replied:** How much time do you spend here? You're really surprised by this?

Anonymous said: Those who deliver nudes deserve no harm

> **Anonymous replied:** hahahahahahaha you must be new here. "for the lulz."*

Anonymous said: I don't wanna be a whiteknight, but already being one, I wonder why /b/ does this. She provided tits and shit, yet "we" do this to her. Internet hate machine at its best.

> **Anonymous replied:** /b/ camwhoring: 2004–2013. R.I.P. Thanks.
>
> > **Anonymous replied:** The amazing thing to me is how you guys never shut up about how "if u keep doxing them we wont have any camwhores left :(." notice that you've been saying this for roughly a decade.

* "Lulz" derives from "lol," or "laugh out loud."

> **Anonymous said:** Anyway here is a list of all her Facebook friends. You can message friends, and all their own friends, so that anyone with a slight connection to sarah via friend of friend knows
>
> > **Anonymous replied:** So has somebody started messaging her friends and family or can I begin with it?
> >
> > > **Anonymous replied:** Assume no one else has, because anyone else who responds might be a whiteknight looking to make you think that someone else was already sending the pics out.
> > >
> > > > **Anonymous replied:** gogogo

One user created a fake Facebook account, put together a collage of Sarah's pictures, and began sending them to Sarah's family and friends with a short message: "Hey, do you know Sarah? The poor little sweetie has done some really bad things. So you know, here are the pictures she's posted on the internet for everyone to see." Within a few minutes, almost everyone in Sarah's social media network had been sent the photographs.

> **Anonymous said:** [xxxxx] is her fone number—confirmed.
>
> > **Anonymous replied:** Just called her, she is crying. She sounded like a sad sad sobbing whale.

Anonymous replied: Is anyone else continuously calling?

This was what /b/ calls a "life ruin": cyberbullying intended, as its name suggests, to result in long-term, sustained distress. It's not the first time that /b/ has doxed camgirls. One elated participant celebrated the victory by creating another thread to share stories and screen grabs of dozens of other "classic" life ruins, posting photographs of a girl whose Facebook account had been hacked, her password changed, and the explicit pictures she'd posted on /b/ shared on her timeline.

Anonymous said: I feel kinda bad for her. She was hot and shit, also cute. Too bad she was dumb enough to leak her name and whatnot. Oh, well. Shit happens.

Anonymous replied: If was clever she would have g[ot] t[he] f[uck] o[ut] she didnt, therefore she deserves the consequences

Anonymous replied: I don't give a shit what happens either. Bitch was camwhoring while she had a boyfriend.

The operation took under an hour. Soon, the thread had vanished, and Sarah was forgotten.

Doxing camgirls is only one of a growing number of ways that people abuse, intimidate, provoke, anger, or "troll" others online. Celebrities, journalists, politicians, sportspeople, academics—indeed, almost anyone in the public eye, or with a large following online—regularly receive insults, inflammatory comments, and threats from complete strangers. In 2011, a British man, Sean Duffy, was imprisoned after making offensive remarks on Facebook, including a post mocking a fifteen-year-old who'd committed suicide. When the UK journalist Caroline Criado-Perez and others succeeded in a campaign to get Jane Austen featured on the new ten-pound note in 2013, she was bombarded with abusive messages from anonymous Twitter users, culminating in bomb and death threats deemed serious enough for the police to advise she move to a safe house. When her father Robin Williams died in 2014, Zelda Williams left Twitter and Instagram after being sent endless insults, horrific photos, and abusive comments. It sometimes feels that anyone, especially women, who puts their head above the internet parapet is targeted.

Some form of trolling takes place on almost every online space. YouTube, Facebook, and Twitter all have their own species of troll, each evolved to fit their environment, like Darwin's finches. MySpace trolls have a register and tone perfectly adapted to upset aspiring teenage musicians. Amateur pornography websites are populated with trolls who know precisely how to offend exhibitionists. The "comment" sections on reputable news sites are routinely bursting with insults.

Over the last five years, there seems to have been a dramatic increase in this type of behavior. In 2007, 498 people in England and Wales were convicted of using an electronic device to send messages that were "grossly offensive, indecent, obscene, or of a menacing character." By 2012, that number had risen to 1,423. In a poll of almost 2,000 British adults on the subject, 2 percent said that they had insulted someone, in some form, online—which, when extrapolated, would amount to some one million trolls in the UK alone. According to the Cyberbullying Research Centre, 25 percent of U.S. high school students have been cyberbullied at some point in their lifetime (16 percent of those surveyed admitted having cyberbullied other people). Based on the National Crime Victimisation Survey, 2.2 million U.S. students (grades 6 through 12) experienced cyberbullying in 2011 (the last time the data was available).

"Trolling" has today become shorthand for any nasty or threatening behavior online. But there is much more to trolling than abuse. Zack is in his early thirties. He has been trolling for over a decade. "Trolling is *not* about bullying people," he insists, "it is all about unlocking. Unlocking situations, creating new scenarios, pushing boundaries, trying ideas out, calculating the best way to provoke a reaction. Threatening to rape someone on Twitter is not trolling: that's just threatening to rape someone."

Zack has spent years refining his trolling tactics. His favorite technique, he tells me, is to join a forum, intentionally make basic grammatical or spelling mistakes, wait for someone to insult his writing, and then lock them into an argument about politics. He showed me one recent example that he'd saved on his laptop. Zack had posted what appeared to be an innocuous, poorly written comment on a popular right-wing website, complaining that right-wingers wouldn't be right wing if they read more. An incensed user responded, and then posted a nude picture that Zack had uploaded to an obscure forum using the same pseudonym some time before.

The bait had been taken. Zack hit back immediately:

> You shouldn't deny yourself. If looking at the pics makes you want to touch your penis then just do it . . . if you want I can probably find you some more pictures of my penis—or maybe you'd like some of my ass also? Or if you want we could talk about why regressive ideologies are a bad idea in general and why people who adopt them are likely to have a much harder time in understanding the world than someone who's accepting of progress and social development?

Zack then began posting a series of videos of his penis in various states of arousal interspersed with insults about right-wingers and quotes from Shakespeare and Cervantes. "Prepare to be surprised!" Zack said mischievously, before he showed me the posts.

For Zack, this was a clear win. His critic was silenced by the deluge, which occupied the comments section of the website for several hours. "He was so incapable of a coherent response that he resorted to digging into my posting history for things he thought might shame me—but I'm not easily shamed."

"But what was the point?" I asked him. "I thought you were trying to expose far-right groups?"

"Yeah, and by posting the naked photos, the discussion drew attention from across the site. This is what trolling is all about—creating a scene in order to get more people to think about the issue being raised."

"And do you think you succeeded in doing that?"

There's a short pause. "I dunno, but it was fun. It doesn't really matter if it was otherwise fruitless."

For Zack, trolling is part art, part science, part joke, part political act, but also much more. "Trolling is a culture, it's a way of thinking"—and one, he says, that has existed since the birth of the internet. If I wanted to discover where this apparently modern problem came from, I had to go back to the very beginning.

FINGER

The internet's precursor, the Arpanet, was, until the 1980s, the preserve of a tiny academic and governmental elite. These "Arpanauts," however, found that they enjoyed chatting as much as exchanging data sets. Within four years of its creation the Arpanet's TALK function (originally designed as a small add-on to accompany the transfer of research, like a Post-it note) was responsible for three quarters of all Arpanet traffic. TALK, which later morphed into electronic mail, or "e"-mail, was revolutionary. Sitting at your computer terminal in your department building, you could suddenly communicate with several people at once, in real time, without ever looking at or speaking to them. The opportunities afforded by this new technology occasionally made the small group of world-class academics behave in strange ways.

One research group, formed in 1976, was responsible for deciding what would be included in an email header. They called themselves the "Header People," and created an unmoderated chat room to discuss the subject. The room became famous (or infamous) for the raucous and aggressive conversations held there. Arguments could flare up over anything. Ken Harrenstien, the academic who set up the group, would later describe them as a "bunch of spirited sluggers, pounding an equine cadaver to smithereens."

In 1979, another team of academics were at work developing a function called "Finger," which would allow users to know what time other users logged on or off the system. Ivor Durham from Carnegie Mellon University proposed a widget to allow users to opt out of Finger, in case they preferred to keep their online activity private. The team debated the merits of both sides, but someone leaked the (internal) discussion to the rest of the Arpanet. Durham was attacked relentlessly and mercilessly by other academics from across the United States, who believed that this compromised the open, transparent nature of the Arpanet.

Most of these academics knew each other, so online arguments were tempered by the risk of bumping into your foe at the next computer science conference. Nevertheless, misunderstanding and righteous indignation spread across the Arpanet. One participant in the Finger episode thought that tongue-in-cheek comments were usually misread on a computer, and proposed that sarcastic remarks made on the Arpanet be suffixed with a new type of punctuation to avoid readers taking them the wrong way: ;-) But even the first emoticon wasn't enough, because users just started slotting them after a sarcastic put-down, which was somehow even more annoying. ("The fing ahole is winking at me as well?!") Worried that the network was quickly becoming an uncivil place, Arpanauts published a "netiquette" guide for newcomers. Satire and humor, it advised, were to be avoided, as "[they are] particularly hard to transmit, and sometimes [come] across as rude and contemptuous."

FLAMING ON BBS

In 1978, in Chicago, Ward Christensen and Randy Suess invented the dial-up Bulletin Board System. With a modem, telephone, and computer, anyone could either set up or connect to a "BBS" and post messages. From the early 1980s onwards, BBS was many people's first experience of life online.

Within a year, insulting strangers on boards became a widely acknowledged and accepted part of BBS. Finger and Header Group disputes were more often than not heated debates between academics. But here, people started joining groups and boards with the sole purpose of starting an argument. This was called "flaming": provoking strangers, disrupting other groups, and creating tension for the fun of it. The best "flames" were well written: subtle, clever, and biting. Good flamers (who would often post under a pseudonym) built a reputation; people would eagerly await their posts, and archive their best lines. This was more than simple nastiness. For many flamers, it was an opportunity to experiment, to push boundaries, and to have their efforts read and appraised. One prominent flamer even published a guide—"Otto's 1985 Guide to Flaming on BBS"—advising potential flamers that being as controversial as possible was "the only way that people will read your opinions." "It is very hard," Otto wrote, "to ignore a board-wide or NET-wide flame war."

Dedicated groups started to appear to discuss how to most effectively flame others. In 1987, one BBS user called Joe Talmadge posted another guide, the "12 Commandments of Flaming," to help flamers old and new develop their style:

> Commandment 12: When in doubt, insult. If you forget the other 11 rules, remember this one. At some point during your wonderful career as a Flamer you will undoubtedly end up in a flame war with someone who is better than you . . . At this point, there's only one

thing to do: INSULT THE DIRTBAG!!! "Oh yeah? Well, your mother does strange things with vegetables."

BBS groups were controlled by a systems operator (sysop), who had the power to invite or ban users, and delete flames before they reached the victim. Often labeled censorsops, they were themselves the targets of a nasty strand of flaming called "abusing." Abusers would torment the sysop with insults, spam, or anything else they could think of. Sometimes abusers and flamers would "crash" a board with bugs, or post links to Trojan viruses disguised as pirated arcade games for unsuspecting users to download. Another trick was to upload messages referring to pirating, in order to direct snooping authorities towards the unsuspecting sysop.

USENET FLAME WARS

Around the same time that the BBS was invented, two academics at Duke University set themselves an even more ambitious task. Tom Truscott and Jim Ellis were aggrieved that the Arpanet was elite and expensive—access cost approximately $100,000 per year—so in 1979 they set up a new network called "Usenet," which, they hoped, anyone could access and use. (Anyone, that is, who had a computer connected to the operating system UNIX, which amounted to very few people.)

Usenet, it can be argued, is the birthplace of the modern troll. Usenetters—a small clutch of academics, students, Arpanauts, and computer nerds—would take a pseudonym and join a "newsgroup" full of strangers. Like BBS, anyone could start a Usenet group, but unlike BBS, the administrators—the people who ran the whole network—had some control over which groups they would allow. The hope of harmony reigning was dashed almost immediately. Usenetters clashed with the haughty Arpanauts over How Things Should Be Done in this new space, with the Arpanauts declaring the new Usenet "trash" to be ignorant and inexperienced. One simple spelling mistake would often instigate a chain reaction, resulting in months of users trading insults and picking apart each other's posts.

Usenetters were a rebellious bunch. In 1987, Usenet administrators forced what became known as the "Great Renaming," categorizing all the haphazard groups into seven "hierarchies." These were: *comp.* (computing), *misc.* (miscellaneous), *news.*, *rec.* (recreation), *sci.* (science), *soc.* (social), and *talk.*—under which users could start their own relevant subgroups. To name the group, you took the main hierarchy name, and then added further categories.* John Gilmore, who would go on to cofound the cypherpunk movement with Tim May and Eric Hughes in 1992, wanted to start a group about drugs, called rec.drugs. His request was turned down by the administrators.

So Gilmore and two experienced Usenetters created their own hierarchy, which would be free of censorship. They called it *alt.*, short for alternative (it was also thought to stand for "anarchists, lunatics, and terrorists"). Flaming became extremely popular on *alt.*, and flamers would take pleasure in being cruel to other users in as creative and imaginative a way as possible. A 1990s Usenet troll called Macon used to respond to flames by posting a single, 1,500-word epic mash-up of creative insults he'd written over the years: "You are the unholy spawn of a bandy-legged hobo and a

* A Usenet hierarchy about this book, for example, might be called "rec .books.darknet."

syphilitic camel. You wear strangely mismatched clothing with oddly placed stains . . ." When, in 1993, a user named Moby asked the group alt.tasteless for advice about how to deal with a pair of cats in heat who were ruining his love life, he received an explosion of maniacal solutions, each more ludicrous than the last: do-it-yourself spaying, execution by handgun, incineration and, perhaps inevitably, sex with the cats.

On both Usenet and BBS, new idioms, rules, and norms were being created. But it was a world that was about to be inundated. The early 1990s saw the number of internet users grow exponentially. And many new users would beeline straight for one of the most active and interesting places online: *alt*. Usenetters, irate at the sudden influx of immigrants, attempted to flush them out. In 1992, in the group alt.folklore.urban, a new type of flaming was mentioned for the first time, targeted at the recent arrivals: trolling. The idea was to "troll for newbies"*: an experienced user would post an urban myth or legend about Usenet in the hope of eliciting a surprised reaction from anyone new, thereby exposing their status. Caught you! The responder would thereafter be mercilessly mocked.

With so many potential targets, flaming and trolling began to spread, and became increasingly sophisticated. Several groups dedicated to trolling were set up in *alt*. In 1999, one user called "Cappy Hamper" listed in the group alt.trolls six different types of trolls: the "straight-up asshole flame troll" ("easy!" explained Cappy Hamper. "Post in alt.skinheads with the header: 'you

* The word "troll" most likely refers to the fishing technique of trailing ("trolling") a baited line, to see what bites, rather than to the mythical cave dweller.

buncha racist asswipers eat dog crap bisquits!'"); the "clueless newbie joke troll"; the "hit, run, and watch troll"; the "confidence" or "tactical troll"; the "creative cross-post troll"; and the "gang troll."

The Meowers were infamous gang trolls. In 1997, a group of Harvard students had joined an abandoned Usenet group called alt.fan.karl-malden.nose to post updates about comings and goings on campus. They then started to mildly flame other Usenet groups, in order, wrote one, "to rile up the stupid people." Matt Bruce, one of the Harvard group, suggested targeting alt.tv.beavis-n-butthead. Users of alt.tv.beavis-n-butthead didn't take kindly to these arrogant students, and started to post back to alt.fan.karl-malden.nose. So did people from other Usenet groups. So much so that the Harvard students abandoned the group, and the Beavis and Butthead invaders took it over, renaming themselves "the Meowers," in mock deference to a Harvard student who, because his initials were C.A.T., signed off his messages with "meow." The Meowers began setting up other Usenet groups (including alt.alien.vampire.flonk.flonk.flonk, alt.non.sequitur, and alt.stupidity), from which they started to invade other groups by posting ridiculous, Monty Pythonesque posts, preventing anyone else from posting or entering into a discussion. This technique, now known as "crap-flooding," is still very popular among trolls. In 1997–98, the Meowers went on a crap-flooding spree, targeting groups across Usenet with what they called their "Usenet performance art." Meowers would also spam individuals who fought back, using anonymous remailers to disguise the address of the sender. The college email system at Boston University was broken by a Meower spam-flood. The campaign lasted for at least two years.

The trolling collective alt.syntax.tactical specialized in the "cross-post" troll. Members would take a genuine post (from a group like alt.smokers, for example) and forward it via an anonymous remailer, with the original email address intact, to a group who, they believed, would not respond kindly (alt.support.non-smokers), sparking an argument between two groups who had no idea that they were, in effect, trolling each other. Alt.syntax.tactical attacks were carefully planned and often involved plants, dummies, and double agents. Trolls like alt.syntax.tactical weren't out for quick wins, but to provoke as large and aggressive a reaction as possible. This is, they argued, what separated the trolls from the flames. A flame was typically just a deluge of insults. Although there was some overlap between the two, a troll was considered to be more careful, subtle, and imaginative: "A troll will hold back, understanding the value of a bigger spank," wrote one anonymous poster to the group alt.troll. And the bigger the "spank" the better:

> Anyone can walk into rec.sport.baseball and say "baseball sucks." It takes unbelievable skill and discipline to cause a PROLONGED flame war. That is what we do. But it can only be done with talent, and numbers to match that talent. We only bring into the fold people who have the knack to use smarts to incite chaos.

Alt.syntax.tactical were explicit in their goals:

- Our names to appear in kill files
- Regulars/Legit people abandon invaded newsgroup
- Receive much hate mail

As trolling spread, so did its reputation. It was at this time that the industry standard response emerged: "Don't feed the trolls!"

A line that spurred many trolls on to increasingly extreme and shocking behavior.

In the late 1990s, trolling took a leap towards the gutter. Trolls of the era had an informal but widely accepted code of conduct: "Trolling is matching wits . . ." wrote one anonymous user in alt .trolls in 1999:

> The contest must be confined to the "level playing field" of Usenet. What someone posts on Usenet is fair game. But real life investigations into what someone posting with their real name does in real life by someone not using their real name (or a common and virtually untraceable one) shouldn't strike anyone as fair.

But the distinction between digital and real was becoming increasingly vague for newer users. Two long-running, infamous episodes put paid to the "real-life" limits. A small disagreement in alt.gossip.celebrities between two posters, Maryanne Kehoe and Jeff Boyd, quickly degenerated into an argument. Kehoe believed that Boyd was spamming the group with pointless messages, and emailed his employers asking for action to be taken against him. The vicious troll, it turned out, was a sensible computer programmer, and had recently become a father. In possibly the longest case of trolling in the history of the net, the American game developer Derek Smart was insulted repeatedly about his (admittedly disappointing) 1996 game *Battlecruiser 3000AD*. "They were your run-of-the-mill antisocial misfits. And when they run into people like me—who doesn't take crap from anyone—well, then everyone cried foul," Smart told me, via email. Arguments in Usenet groups when the game was released spread across the internet and just kept ratcheting up, partly because Smart kept counter-trolling.

"Back in the day," he confessed, "I let this sort of thing get to me." By 2000, most of the comments concerned Smart's personal life and professional credentials, and most were allegedly posted by a man named Bill Huffman, a "self-proclaimed Derekologist," and the manager of a California software company. Smart was also stalked by a sixteen-year-old who claimed to own a gun. Smart applied for restraining orders and filed complaints with a confused police force. The final dispute—concerning a website Huffman had set up—was only settled in 2013.

This niche online world was being subsumed by the newcomers: Usenet codes of conduct about trolling were increasingly meaningless. It was about to get a lot worse.

GNAA AND GOATSE

By the late nineties some feared that Usenet would be ruined by trolling. In the end, innovation killed it off. The internet was becoming more accessible and the speed of downloading (and more importantly, uploading) was slashed, enabling users to post more content online, including pictures and videos. Usenet, like most new and exciting technologies, had become outdated.

At the turn of the millennium, trolls migrated from Usenet to a new breed of irreverent, user-driven, censorship-free sites, that were soon collectively labelled as "Not Safe For Work" (NSFW), and often created by students or teenagers: SomethingAwful.com, Fark.com, and Slashdot.com. Unlike traditional media, these sites were filled with stories, links, suggestions, and comments from their readers. Whatever stories were the most read or shared by users would rise up the ranking system, meaning popularity was driven not by centralized editorial control but by whatever happened to capture the attention of the community. This created—as with many content-driven sectors online—a natural incentive to be outrageous. Stories that were offensive, rude, or bizarre were usually the most popular. Fark had one million unique visitors in its fourth year of existence*: a decent slice of the internet pie in 2000, when only 360 million people in the whole world were online.

The denizens of these new sites adopted and extended the philosophy of their trolling predecessors: abhorrence of censorship—which was thought of as archaic and analog—and the idea that nothing online was to be taken seriously. The humor—which still characterizes a lot of internet culture—was abstract, self-referential, and irreverent.

Trolls pressed offensiveness into the service of this ideology, often in creatively disgusting ways. Goatse is short for "goat-sex." It is also the name of a website set up in 1999. (I don't advise that you search for it.) The home page features a photograph of a naked middle-aged man stretching open his anus. Trolls used the website for "bait-and-switch" pranks: the posting or sending of harmless looking links that actually direct clickers to the Goatse website. This is also known as "shock trolling." In 2000, Goatse links were repeatedly posted on Oprah Winfrey's "Soul Stories" chat board, with misleading accompanying messages: "I've been feeling so down lately, here's a link to a poem I've written." There was an exodus of offended Oprah fans, and at one point the whole

* An impressive number considering that for its first two years the site was nothing more than a photograph of a squirrel with enormous testicles.

board was shut down. The SomethingAwful users behind the prank celebrated this strike against the earnestness that seemed to be spreading across the internet.*

The Gay Nigger Association of America (GNAA) was created in 2002, and typified this sort of extreme trolling. Their opening page featured the following invitation: "Are you GAY? Are you a NIGGER? Are you a GAY NIGGER? If you answered 'Yes' to all of the above questions, then GNAA might be exactly what you're looking for!" The creators of GNAA were reportedly highly skilled programmers,† and dedicated an enormous amount of time to creating and disseminating extremely offensive material, with the aim of upsetting bloggers, celebrities, popular websites, and anyone else the group took against. It would often crap-flood sites—filling chat functions with nonsense, just as the Meowers had done a decade earlier—and hack other popular websites to alter them. GNAA described their purpose as "sowing disruption on the internet" but eventually set up an internet security organization, hacking into sites to demonstrate how susceptible to attack they were. They called it Goatse Security—"exposing gaping holes"—and while members of the group have been investigated by the FBI for various hacking offences, Goatse Security has also identified and fixed a number of security flaws in major internet products and software. Zack was an early admirer of GNAA and Goatse. "People were just so ready to be offended by things like Goatse," he tells me. "It's fun to upset someone who is so ready

* SomethingAwful users continue to attack other boards and forums in a similar way.

† The former "President" of GNAA is considered to be a hacker and troll named "Weev."

be offended. And when they get upset, they prove you're right. It's circular."

DOING IT FOR THE LULZ

In some ways Zack, GNAA, and other NSFW trolls felt it was "their" internet that was being invaded by marketers, celebrities, big business, the authorities, and legions of ordinary people, in the same way Usenetters felt inundated in 1993. People outside of the tribe, and all of them taking everything so *seriously*. Out of this milieu came Christopher Poole, a fourteen-year-old fan of SomethingAwful from New York, who had found a Japanese image-sharing website called Futaba that allowed users to post about anything, anonymously. NSFW sites were exciting and bold—but participants were often identifiable, and sites were frequently moderated. The anonymous Futaba users were wildly creative, highly offensive, and uncontrollable. The website was notorious in Japan for gory fiction about students slaughtering teachers, anime porn, and much besides, causing general moral outrage. Futaba's web address was www.2chan.net, in tribute to the similarly outrageous website 2channel, so when Poole decided to set up an English-language equivalent in 2003, he called it 4chan: "its [sic] TWO TIMES THE CHAN MOTHERFUCK!" he posted under the pseudonym "moot."

Zack joined immediately: "We were trying to carve out our own space, our own part of the internet." The quasi-enforced anonymity made /b/ a natural home for trolls. Trolling in /b/ is widespread and extremely varied, with dozens of different trolling categories. The hacktivist collective Anonymous were almost all committed /b/tards, and used the site to plan and coordinate their "operations." The group's first major action was called Project Chanology, directed against the Church of Scientology after the Church tried to remove embarrassing videos of Tom Cruise from the net. Although the message was a genuine one—about censorship and transparency—alongside the serious demonstrations and computer hacks were endless prank phone calls to the Scientology hotline, 4chan-inspired placards, and hundreds of black faxes.*

Enforced anonymity, the competitive urge to outdo your fellow users and a determination to push offensiveness in the name of a vague anticensorship ideology are all wrapped up in a /b/ trolling catchphrase: "I did it for the lulz"—a phrase employed to justify anything and everything where the chief motivation is to generate a laugh at someone else's expense. The problem, as Zack explains, is that "lulz" are a bit like a drug: you need a bigger and bigger hit to keep the feeling going. Trolling can quickly spiral out of control. The popular social networking and news-sharing site Reddit once hosted a group called Game of Trolls. Its rules were simple: if you successfully upset someone on Reddit without them realizing they were being trolled, you won a point. If you were identified as a troll, you lost a point. The highest scorers were listed on a leaderboard. One user visited a popular subreddit and posted an invented story about the problems he was having with a coworker. The same user then replied as the coworker in question, demanding an apology, and explaining that he had difficulty making friends. Redditors believed the story, and some even offered

* A "black fax" is a fax of a black page, sent to waste the ink of the recipient's fax machine.

to send flowers to the abused colleague. The group had been successfully trolled. "It was glorious," recalled a witness. Game of Trolls was eventually banned by Reddit; a highly unusual step for the otherwise liberal site, but testament to the pervasiveness and persistence of the Reddit trolls.

The competition to insult and offend, by any means necessary, can often lead to shocking extremes. In 2006, Mitchell Henderson, a fifteen-year-old from Minnesota, committed suicide by shooting himself with his parents' rifle. Mitchell's classmates created a virtual memorial for him on MySpace and wrote a short eulogy, which included repeated references to Mitchell as "an hero": "he was an hero to take that shot, to leave us all behind. God do we wish we could take it back." The combination of a grammatical error with the contention that committing suicide was a "heroic" act caused great hilarity on 4chan. After learning that Mitchell had lost an iPod shortly before killing himself, the /b/tards created photoshopped images of Mitchell and his lost device. One even took a photo of an iPod on Mitchell's grave, and sent it to his bereaved parents. For almost two years after his death, they received anonymous phone calls from people claiming to have found Mitchell's iPod.

MEETING THE TROLLS

Finding genuine trolls is difficult. Many use proxy servers to mask their IP addresses and most have dozens of accounts with different names for each platform they use. If they are banned or blocked by a particular site, they'll just rejoin under a new name. But like the Meowers, today's trolls enjoy spending time with other trolls.

A lot of the worst trolling is coordinated from hidden or secret channels and chat rooms.

Zack agreed to show me one of his hideouts, inviting me in to a secret channel he's been frequenting for over two years. It's a private group on a well-known social media page, "a pirate base for trolls," he told me. The main group page—the one the typical user sees—is a series of pictures of people masturbating. "That's a façade," says Zack, "to keep away the dullards." To get to the real action, you have to be invited to join as a moderator by an existing moderator, granting you access to the group's internal mailing system. Inside, the pace is frantic: every day there are constant, lengthy, and hilarious arguments and discussions that draw in up to twenty moderators at once, some of whom know each other and some of whom do not. Everyone uses a fake name, because everyone has been banned from the site before, so I didn't stand out—I could have been anyone. Everyone is trolling each other endlessly here, and most of the messages are very funny, and extremely sharp. According to Zack, at least two of the contributors are university professors. It feels like a training ground for trolls, a place to go to try out new tactics and battle others without too much damage being done. A place to come and relax, to wind down a little with some peers.

While I was there, an infamous troll became a moderator too. Zack explained to me that this particular troll described himself as an "incel," short for an "involuntary celibate." This troll was well known in trolling circles for having run a blog in which he argued at length that the government owes him a woman to have sex with; and he boasted that he became so desperate that he once tried to have sex with his own mother. When he pushed his line on the group—that he should be able to have sex with anyone he wants, that the government should help him do so, that all girls are sluts anyway—no one could quite work out if he was trolling them or not. They were all fascinated, though, and started probing back, counter-trolling:

> Hey incel are you homophobic? like just for example pretend this was a real room we're in and me and _______ started kissing, how would you feel about that? if it became really passionate and i was squeezing his soft little bum as i pushed my tongue deep down his neck? would you have opinions on that?
>
> No
>
> Hey incel—is your mother pretty? how many out of ten? is she a 7+ I'm just curious if your need for a girl to be hot as fuck (else you abuse them) extended to your mother . . .
>
> Um, yeah . . . I can only laugh at ya

The other trolls in the group appeared to be sizing him up, searching for weaknesses. This is called "trolls trolling trolls": when no one is really sure who's trolling whom. It's not about winning or losing, more like sparring.

Old Holborn has been called Britain's "vilest troll" by the *Daily Mail* newspaper for his endless online abuse, including his attacks on the families of the ninety-six Liverpool soccer fans who died in a stadium collapse in 1989. He tweets and blogs constantly, hiding his face behind a Guy Fawkes mask. Without it, he's not quite as intimidating: a well-dressed, fast-talking middle-aged man from Essex, a county that border London. He's a successful computer programmer and recruitment specialist, he tells me. "You could call me a gobshite," he explains over coffee. "Always have been. I'm very antiauthoritarian." He's more than that, he's a minarchist—someone who believes in the smallest possible government. "We just need someone there to protect private property. Everything else, we can work out ourselves." He sums up his world view: "the government should just leave us alone." Trolling is his way of causing trouble for the system: "I want to be the itch, the grain of sand in the machine." In 2010, he stood for the UK Parliament in Cambridge, wearing his mask and frustrating the Electoral Commission by changing his name to Old Holborn by deed poll. Around the same time, he marched into a police station in Manchester wearing the mask and carrying a suitcase full of cash to post bail for a pub owner who'd refused to enforce the British government's 2007 ban on smoking in enclosed public spaces. This, he says, is also trolling.

It's hard to see what insulting the families of victims of the soccer stadium disaster on Twitter has to do with minarchy. But there is a link. To live in Old Holborn's libertarian stateless utopia, people need to be tough and independent, and take responsibility for their actions. He fears a silent and obedient society, and says that one where everyone is easily offended will lead to self-censorship. He sees it as his role to prod and probe the boundaries of offensiveness to keep society alert. He targeted the Liverpudlians, he says, to "prove" that they suffered from victimhood syndrome—and for Old Holborn, just like the Usenet trolls, the unit of success was the reaction. "I would be controversial, just to show that they loved feeling like the victims. The response was phenomenal: they threatened to burn down my office, my house, and rape my children. Haha! I was right! They proved I was right!"

As a result, though, he was doxed and—soon after we met—he'd moved to southern Bulgaria to, in his words, "cause trouble full-time" from there. "I'm the good guy!" he shouted in the cafe. "I'm the one exposing the hypocrisy. I'm the one trying to make society freer!"

THE TRUTH ABOUT TROLLS

In the 1980s and '90s, as a growing number of people went online, psychologists became interested in how computers were changing our thoughts and behavior. In 1990, the American lawyer and author Mike Godwin proposed a natural law of Usenet behavior: "As an online discussion grows longer, the probability of a comparison involving the Nazis or Hitler approaches 1." In short, the more you talk online, the more likely you'll be nasty; talk long enough, and it's a certainty. (Godwin's Law can easily be observed today on the pages of most newspapers' online comment boards.) In 2001, John Suler's famous Online Disinhibition Effect put forward a reason why. It listed six factors that, Suler claimed, allowed users of the internet to ignore the social rules and norms at play offline.* He argues that because we don't know or see the people we are speaking to (and they don't know or see us), be-

* Suler's six factors are dissociative anonymity (my actions can't be attributed to my person), invisibility (nobody can tell what I look like, or judge my tone), asynchronicity (my actions do not occur in real time), solipsistic introjection (I can't see these people; I have to guess at who they are and their intent), dissociative imagination (this is not the real world; these are not real people), and minimization of authority (there are no authority figures here, I can act freely).

cause communication is instant, seemingly without rules or accountability, and because it all takes place in what feels like an alternative reality, we do things we wouldn't in real life. Suler call this "toxic disinhibition." According to other academic studies, between 65 and 93 percent of human communication is nonverbal: facial expression, tone, body movement. Put very simply, our brain has evolved over millions of years to subconsciously spot these cues so we can better read and empathize with each other. Communicating via computers removes these cues, making communication abstract and anchorless. Or, as the web comic Penny Arcade has it: "The Greater Internet Fuck-wad Theory": "normal person + anonymity + audience = total fuckwad."

The easiest way to deal with the trolls is to remove their anonymity, to force websites or platforms to insist that everyone log in under their real names. That wouldn't stop online nastiness entirely of course, but it would at least make trolls a little more accountable for their actions, and perhaps encourage them to hesitate before abusing others. But removing anonymity online has its drawbacks. Anonymity is not a modern invention designed to protect trolls. It also allows people to be honest and open and invisible when there are good reasons to. We dispense with that at our peril.

Get rid of trolling and we might lose something else, too. The line between criminality, threats, offensiveness, and satire is another very fine one. Trolls like Old Holborn do occasionally cast a satirical eye on society's self-importance, exposing the absurdity of modern life, moral panics, or our histrionic twenty-four-hour news culture. One branch of trolls, called "RIP memorial trolls," target people who post messages to online memorial pages of the recently deceased. According to Whitney Phillips, an academic who wrote her PhD on trolls, they usually target what they call "grief tourists": users who have no real-life connection to the victim and who could not possibly be in mourning. The trolls themselves claim that grief tourists are shrill, disingenuous, and wholly deserving targets. The Gay Niggers Association of America frequently posts ridiculous news stories in the hope that lazy journalists will repeat them. They often do: a GNAA story alleging that African Americans were looting people's homes during Hurricane Sandy in order to steal domestic pets was widely reported by mainstream media outlets. Within the trolling community, the undisputed champions of trolling "in real life" are considered to be U.S. comedian Stephen Colbert and British comedy writer Chris Morris, both famous for puncturing the inflated egos of politicians and celebrities.

Zack claims that his work has value and purpose too—"trolling in the public interest" to expose hypocrisy and stupidity in society. He has even created his own complicated religion, which he's spent years pulling together, simply to use as a trolling tool. He calls it "autodidactic time travel pragmatism," a mixture of absurd humor, physics, and fragments borrowed from other religions. He uses it to troll religious and political groups. "It's the tried and tested hazing technique of presenting someone with something that is impossible to know whether or not to take seriously—impossible to know where the joke ends and the seriousness begins." It's a clever tactic, and—to my surprise—one with significant ramifications for contemporary theological debate.*

While many trolls are simply bored teenagers trying to cause a

* You have been trolled.

little trouble, the serious trolls seem to broadly follow a libertarian ideology, and believe that part of living in a free society is accepting that no idea is beyond being challenged or ridiculed, and that nothing is more stifling to free expression than being afraid to upset or offend. Trolls have existed just as long as networked computing, which surely says something about the need many of us have to explore the darker sides of our nature. Every troll I've spoken to says what they do is natural, a human need to push a boundary simply because it's there.

The problem with a boundary-pushing philosophy is that it can be used to justify bullying and threatening people with no regard for the consequences. When I asked Zack if he's ever gone too far he nods, "Yeah, I guess there were a few people who I hounded so bad that they left the internet. One had a mental breakdown." Does he feel responsible? "At the time, I didn't—we all knew what we were doing. Although now I'm less sure." Old Holborn is more resolute: "I pick my targets carefully. They always deserve it." But the powerful and the rich are not always the targets. Too often it is the weak, the newcomers like Sarah, who are the easiest to attack. Anonymous users on /b/ pick on camgirls because their pictures and threads are wildly popular: far more than the normal /b/tard thread. Ultimately Old Holborn agrees with /b/: "Would you go post photos of yourself and put them on the internet? So why did she do it? It's not about teaching her a lesson, but she has to be responsible." Zack is uneasy about Sarah's case, but finally concludes, "Well, she probably shouldn't have done that, although she didn't deserve the consequences."

To me, the Sarah dox just felt like crude nastiness. The perpetrators made a limp effort at justifying it: "That silly bitch may have learnt the most important lesson of her young life tonight: that posting pictures of your naked body on the internet is a monumentally bad idea." I'm sure she did learn a painful lesson, but that was only a side effect of the "life ruin":

> **Anonymous said:** I'm a moral fag
> I see no problem in doxing sarah's ass
> It's for the lulz
> At the top of the tree of life there isn't love: there is lulz

Whatever their motives, and even at their worst, perhaps we can learn something from trolls. Trolling is a very broad church, ranging from /b/ bullies to amateur philosophers, from the mildly offensive to the illegal. An increasing desire for digital affirmation is leading more of us to share our most intimate and personal lives online, often with complete strangers. What we like, what we think, where we're going. The more we invest of ourselves online and the more ready we are to be offended, the more there is for trolls to feed on. And despite the increasing policing of social media sites, trolling is not going anywhere. It has been a central feature of online life since the mid-1970s, evolving and mutating from an unexpected offshoot of electronic communication within a niche community to an almost mainstream phenomenon. For people like Zack, the degeneration of trolling from creative art to random threats and bullying is frustrating. But that won't stop him.

Whether we like it or not, trolling is a feature of the online world today. As we all live more of our lives online, trolls might help us to recognize some of the dangers of doing so, make us a little more careful, and a little more thick-skinned. One day, we might even thank them for it.

EPILOGUE

Four days after Sarah's ordeal, another /b/ camgirl was doxed, with photos sent to all her family members, her employer, and her boyfriend: "Do you know your girlfriend posts pictures of her tits on the internet? You can see them here _____"

"Another day, another harsh reality," wrote one.

"She'll be back," replied another.

2

THE LONE WOLF

I first met Paul in a working men's club in a town in the north of England one cold autumn evening. He looked young, with a handsome face, short dark hair, and tattoos that climbed up his neck. He was good company: polite, attentive, and quick to laugh. In short, Paul and I got on very well. Until, that is, talk turned to politics. "Just think of the beauty that will die, Jamie," he explained. "What do you think the world will be like under black or Paki or brown rule? Can you imagine it? When we're down to the last thousand whites, I hope one of them scorches the fucking earth, and everything on it."

Paul is a one-man political party, a propaganda machine. He spends all day, every day, trying to spark a racial awakening among white Britons. He runs a popular blog about ethnocentrism and White Pride, and produces and posts videos attacking minority groups. He opens his laptop and shows me his recent activity: a heated debate with members of a left-wing political group; messages of support for the Greek Golden Dawn party; communications with white supremacists in the United States. He loads up his Facebook and Twitter pages. Thousands of people, from all over the world, follow Paul's breathless output on social media. Online, he has found a community who share his beliefs and appreciate his posts. He has also attracted an equally vocal group dedicated to opposing his views and taking him offline. He lives in a one-dimensional world of friends and enemies, right and wrong—and one where he has been spending increasing amounts of time. The digital Paul is a dynamic, aggressive, and prominent advocate of the White Pride movement. The real Paul is an unemployed thirty-something who lives alone in a small house.

On the train home after one of our interviews, I texted him a note of thanks. As usual, he replied immediately: "You're very welcome Jamie :-) Have a safe journey back. PS I really enjoyed it." But unlike previous meetings, very soon afterwards our correspondence slowed. The usually vocal Paul went quiet. His social media activity stopped. Maybe, I allowed myself to hope, our meetings had made a difference? Or maybe the police had finally worked out who he was? Perhaps something worse?

A NEW PLATFORM

Paul is not alone in finding the internet a perfect place to spread his message. It has become a vital platform for political groups around the world. From Barack Obama's Facebook electioneering in the United States, to the Occupy movement's flash mobs, to the Italian comedian-cum-politician Beppe Grillo's digital reach, the battle for ideas, influence, and impact is moving online. But the same techniques are being used by extremist political movements to spread a message of hate and recruit a new generation of supporters. Most recently, the so called "Islamic State" (formerly ISIS) have used social media with devastating effect: their propaganda stays online and reaches millions. The way the internet might make it easier for terrorist groups to "radicalize" young people into supporting extremist ideology has governments around the world worried. In February 2015, President Obama convened a "Countering Violent Extremism Summit" at the White House in order to better fight extremist propaganda. The main focus of discussion? The internet.

Over the last decade, Paul and thousands of people like him have eschewed the traditional stomping grounds of the nationalist movement in favor of Facebook, Twitter, and YouTube. They were among the first political groups to do so. Extremist organizations, denied a platform on mainstream media and unable to propound their beliefs in public, were particularly attuned to the opportunities that new outlets and platforms gave them. In the 1980s and '90s, for example, the American white supremacist organizations Stormfront and the Aryan Brotherhood created and maintained popular support groups on Usenet and Bulletin Board Systems. (In fact, Stormfront started life as a website.) Blood and Honor, the epicenter of the extreme neo-Nazi music scene, has dozens of open YouTube pages and closed online discussion forums. Stormfront's website—stormfront.org—hosts a long-standing forum, which has close to 300,000 members, who between them have posted close to ten million messages. Twitter is especially popular among neo-Nazis, who will often take a username including the numbers 14 and 88. Fourteen refers to the "14 words" ("we must secure the existence of our people and a future for white children"), while 88 refers to the eighth letter of the alphabet.

HH: *Heil Hitler.* According to researchers at King's College London, neo-Nazis use Twitter not only for disseminating ideas and sharing propaganda, but also for maintaining a coherent sense of self-identity. "The House of Rothschild must be destroyed if we are to save our race! 14/88. Sieg Heil!" posts one United States–based user I find after a cursory search. Some nationalists find children's chat rooms or innocuous-sounding Yahoo! groups in which to meet. Anglo-Saxon history discussion boards are particularly popular among English nationalists, where hundreds of users with names like "Aethelred" and "Harold" discuss how to establish a purer, whiter England. In early 2007, supporters of the French nationalist party the Front National became the first European political party to set up a political office in the virtual world "Second Life," prompting a wave of online protests from other users. The same year, its xenophobic avatars visited a virtual mosque, sat on the virtual Koran, and posted anti-Semitic slogans, before activating a hacker script that automatically ejected everyone from the building. The Jewish human rights organization the Simon Wiesenthal Center estimates that as of 2013 there were 20,000 active "hate" websites, social network groups, and forums online, based all over the world. The number is growing every year. The online world has become a haven for racists and nationalists, giving political extremists an opportunity to voice their opinions, share ideas, and recruit supporters.

Nick Lowles, director of the campaign group Hope Not Hate, has been working for antifascist groups since the mid-nineties. Nick tells me the internet "has given the ordinary person access to far-right groups in a way that was impossible a decade ago." It is also changing the demographic of the typical nationalist, explains

Nick. It's no longer the jackbooted skinhead he used to target. The modern nationalist is young, time-rich, technologically literate—able to quickly and easily connect virtually to like-minded people around the world. People like Paul.

The most infamous of this new breed of online extremists is Anders Behring Breivik, the right-wing extremist who killed seventy-seven people in a terrorist attack in Norway in July 2011. After leaving school he started to work in customer service, but his talent for computer programming led him to start his own computer programming business. By his early twenties, the young Breivik was spending hours each day reading online blogs and articles about the imminent end of the white race and the threat of "cultural Marxism" to European culture. He became convinced that Islam was taking over Europe, and that violent resistance was the only way to curb its rise.

In the years leading up to his attacks he wrote, under the pseudonym Andrew Berwick, a 1,516-page manifesto titled *2083: A European Declaration of Independence.* It is part memoir, part practical manual for what he saw as a coming race war. Large chunks of it were copied and pasted from the net (he later admitted in court that he'd taken a lot of it from Wikipedia in particular), from sources as eclectic as the seventeenth-century philosopher Thomas Hobbes and the American anti-Islamist commentator Robert Spencer, whose work is cited at length.

Unusually for a terrorist attack of this size and scale, Norwegian security services believe Breivik acted entirely alone—a "lone wolf" with no accomplices or coconspirators. The term was popularized by the American white supremacist Tom Metzger in the 1990s, when he advised fellow neo-Nazis committed to violent action to act alone, in order to evade detection. According to Jeffrey D. Simon, author of *Lone Wolf Terrorism: Understanding the Growing Threat*, the lone wolf is "the most innovative, most creative and most dangerous" type of terrorist. Lone wolves aren't restricted by ideology or hierarchy, and don't need to worry about alienating their group or organization.

More importantly, their lack of communication can make them difficult to identify. In Simon's view, the wealth of easy-to-access information facilitates the rise of lone wolves. The number of lone wolf cases has increased steadily over the last decade, including the Islamist Major Nidal Malik Hasan who murdered thirteen fellow soldiers at the Fort Hood army base in Texas in November 2009, in protest, it is believed, of the wars in Iraq and Afghanistan.

Breivik was a lone wolf: but he did have a network. He believed social media—especially Facebook—would help the white "resistance movements" fight back against the multiculturalism he detested, because it offered new opportunities to push propaganda and connect with like-minded individuals around the world. He wanted to distribute his manifesto to sympathizers, whom he hoped would use it to further the nationalist cause, and perhaps even imitate him. So, over the course of two years, he painstakingly created a vast virtual community, using two Facebook accounts to connect to thousands of fellow extremists across Europe. In *2083*, he documents the long hours spent on the monotonous, but important, task of finding them:

> I'm using Facebook to target various nationalist-related groups and inviting every single member [to become my Facebook friend] . . . aaaaarrrrggh:/ It's driving me nuts, lol . . . I've been doing this for 60

straight days, 3–4 hours a day . . . God, I wouldn't have imagined it was going to be this f...... boring :D

Breivik was after email addresses—which he gathered by requesting and becoming friends with Facebook users.

By early 2011, Breivik had thousands of Facebook friends and contributed to several online blogs, including the right-wing Norwegian site document.no, where he commented on a number of articles criticizing Islam. According to the Southern Poverty Law Center, Breivik had also been a registered member of Stormfront since October 2008, using the handle "year2183." By June 2011, he'd farmed 8,000 "high quality" email addresses; "Ofc, it's a quite tedious task," he admitted, "but then again, I can't think of a more efficient way to get in direct touch with nationalists in all European countries."

Breivik saw opportunities everywhere online. Wikipedia, he suggested, was a good place to nudge public opinion through carefully editing pages. He played the online shooting game *Call of Duty* to hone his shooting skills (he was also a fan of online games like *World of Warcraft*) and advised fellow resistance fighters to use the anonymous browser Tor to evade government detection. Towards the end of *2083*, Breivik made a plea to all patriots to "create a nice website, a blog and establish a nice-looking Facebook page . . . to market the organization."

Suddenly in July 2011 the usually vocal Breivik went quiet. His social media activity stopped. On the morning of July 22, he posted a YouTube video urging comrades to embrace martyrdom. A few hours later he emailed his manifesto to over 1,000 of the addresses he'd harvested from Facebook. At 3:25 p.m., he detonated his homemade bomb outside government buildings in central Oslo, killing eight, before travelling to Utøya Island, where he shot and murdered a further sixty-nine activists from the Norwegian Labour Party who were attending a youth camp.

Exactly who received *2083* remains a mystery. Around 250 people in the UK were sent a copy. Some of them were supporters of a very popular English Facebook page that Breivik had "liked" using a pseudonym in early 2010, and that he praised in his manifesto. This is where Paul's journey began.

E-E-EDL

The English Defence League is characteristic of a new wave of loosely related nationalist movements growing across Europe. Its ideology is difficult to pin down, but it combines a concern that large-scale immigration—especially from Muslim countries—is destroying national identity with a belief that the elite, out-of-touch liberal establishment don't know or care what this is doing to ordinary people. It is usually ostensibly non-racist, claiming to support equality, democracy, freedom, and traditional British (or sometimes Christian) culture. Above all, it believes Islamic and British values are incompatible.

Since the Second World War, membership of formal political parties in the UK has fallen from over three million in the 1950s to under half a million in 2013. Unlike a traditional political party, membership to the EDL is open to all: it demands no money, energy, or time. By 2012, the EDL had become one of the most recognizable street movements in the UK. Supporters had held hundreds of demonstrations across the country, and joined the Facebook group in their thousands. For a nationalist group, its rapid success was unexpected and unprecedented. At its peak in 1973, when the UK was gripped by fears about immigration, the extreme right-wing party the National Front had approximately 14,000 members. The British National Party's peak membership, in 2009, was roughly the same. It took these parties years of concerted campaigning to accumulate these numbers. It took the EDL months. It has local branches in every region of the country and demonstrations, protests, and events take place every month as supporters flit easily between the online and offline world. Its size and scale belie its humble origins—a simple Facebook account.

In March 2009, a small number of radical Islamists from Luton, a small town just outside London, announced they were planning to stage a protest against the British military presence in Iraq and Afghanistan at the homecoming parade of the Royal Anglian Regiment. Stephen Yaxley-Lennon—who now goes by the name Tommy Robinson—read about the protest, and knew of the group, who regularly handed out leaflets near the tanning salon he owns in Luton town center. Although Tommy had briefly been a member of the British National Party, he wasn't particularly interested in politics, but was incensed by the planned protest. Together with some friends, Tommy decided to oppose the group and support the soldiers, to show the world "that Luton wasn't overrun by Islamic extremists."

At that first demonstration—which comprised a few dozen people—there was a skirmish, and the story was picked up by local newspapers. Tommy and his friends decided to create a new group to disrupt meetings and the recruitment efforts of Luton Islamist organizations. He phoned the few contacts he had in small patriotic and nationalists groups, including the United British Alliance. They called themselves the "United People of Luton," and staged a second, larger demonstration in June 2009. It attracted hundreds of people, and resulted in clashes with the police, and nine arrests.*

Tommy had paid a cameraman £450 to make a short video of the day, which he then posted on YouTube. "This time," he tells me in a pub just over the road from New Scotland Yard, "I went on to all the online football forums, chat forums, posting it." He instantly began to receive messages of support from across Britain, and beyond. A dozen or so members of the nascent movement met in a pub shortly afterwards to discuss the future. They decided to create an online organization—one with an international reach. Together with a friend, Tommy signed in to Facebook to create a new group, calling it the English Defence League.

LIKE

As a recruiting tool and organizational platform for a fledgling nationalist movement with no money, and little support, Facebook was unsurpassable. It opened up a whole new world. Within a few hours of the group being created, hundreds had signed up. "It went mental," Tommy recalls. "Lads from all over the country were joining." It was a cheap and effective way to recruit new people,

* The following month another demo was held by the UBA and UPL, and turned nasty—with a number of arrests. Chants included "U—U—U—BA," and "What's it like to wear a dress?" The UBA put forward a spokesperson named "Wayne King." It was, in fact, Tommy. "I picked the name for a laugh," Tommy told me, "so when Victoria Derbyshire [a BBC radio presenter known for her husky voice] introduced me on the radio, it sounded like wanking!"

communicate information about upcoming demonstrations and share stories and photos of previous protests. "Queen Lareefer"—a female supporter of the EDL in her late twenties—was initially attracted to the EDL Facebook page when she spotted a friend sharing a link to a discussion about a recent news item: "People were talking about the poppy burning on Facebook and I saw that someone had liked the EDL page, so I went on to it, I liked the page, I made a comment, someone replied, and I got talking." She went on her first demonstration the following month.

By the end of 2010, the EDL had used Facebook to organize around fifty street demonstrations across the country—some with as many as 2,000 participants. Although the group's site states their commitment to peaceful demonstrations, their meetings were often accompanied by drunkenness, violence, antisocial behavior, Islamophobic chants, and arrests—frequently involving clashes with the left-wing street movement Unite Against Fascism. But the group's reputation grew, and so too did the media coverage, which in turn drove more people to the Facebook page and to the EDL's website.

While Tommy was charging up and down the country on monthly demonstrations, Paul was drifting: taking drugs and partying, for the most part. One day in the summer of 2010, he received a Facebook update when a friend "liked" the English Defence League's page. "I'd never heard of them until then," he tells me. "But something about the name piqued my interest." He wanted to learn more, so he too clicked "like," and began to receive daily updates about this new movement.

Like Paul, anyone could simply click to join the Facebook group, and click to leave just as easily. But more were joining than were leaving, and many wanted to do more than "like." Soon enthusiastic supporters were setting up their own EDL pages and groups, keen to start local chapters and arrange their own demonstrations. Although the leadership decided to impose a more formal structure on the rapidly expanding organization in 2010—dividing up the management and administration along area-based and thematic divisions—it remained a uniquely loose, decentralized, and flexible movement.

But this type of membership model has downsides. By late 2012, initial enthusiasm among the rank and file waned, as members realized long-lasting political change takes more than online chatter and weekend demonstrations. And with such a changeable hierarchy, the group quickly split into several warring subgroups and factions. By early 2013, the EDL was on the verge of implosion. Tommy (who by this point had spent time in jail for breaching bail conditions that forbade him from attending demonstrations) was exhausted, inundated with death threats, and ready to quit. Then, on the morning of May 22, 2013, a British soldier called Lee Rigby was murdered by two radical Islamists in broad daylight in the middle of a busy south London street. In the weeks that followed, the group's online support increased dramatically, and Tommy found himself all over the mainstream media. He couldn't leave.

ADMINS AND MODS

Soon after Paul joined the EDL's Facebook page, he began interacting with other members and posting comments.* His frequent, articulate, and aggressive contributions were getting

* The EDL's Facebook pages are usually "public," allowing any users to post there.

him noticed by the senior members who ran the page. A few weeks later he was invited to join a secretive Facebook group of hard-core EDL members, operating under a cover name. Shortly after that, he was asked to become a moderator or "mod" of a page dedicated to outing Islamist extremists. It was a big step up for Paul. Before he knew it, he was a part of something.

Whether it's a closed or open forum, someone needs to control the chaos and regulate the tone of conversations. It's an important role, because you have power to ban users, and delete or edit other people's posts. By early 2012, over 1,000 people were part of the group Paul administered. Not only did Paul have a voice and a platform, but also an increasing level of power and responsibility. "I loved it," he said. "I'd be on there for hours—posting, monitoring, editing."

Running Facebook groups and Twitter accounts is an extremely important position in a nationalist group. When Lee Rigby was murdered, Tommy immediately contacted the members who ran the group's social media pages. He asked the administrator of the EDL Twitter account to put out a call to arms. At around 6:30 p.m., an announcement was made:

> EDL leader Tommy Robinson on way to Woolwich now, Take to the streets peeps ENOUGH IS ENOUGH.

Hundreds of people retweeted the message, spreading it to thousands of others. EDL supporters quickly started to gather in southeast London.

The EDL's Twitter administrator is a polite sixteen-year-old girl named Becky. As of writing, approximately 35,000 people follow the regular updates about important stories, information about demonstrations, propaganda, and encouragement she posts on the EDL's official feed. Like Paul, she was put in charge when the former EDL Twitter admin noticed she was regularly posting relevant messages and links from a personal account and invited her to help out. After "proving herself" while another admin watched over her, she was made a permanent admin. It is a busy and important job, she explains: "Sometimes I go on it from the moment I get up, till when I go to bed." Even when she is out with her friends, she's still tweeting: "But it doesn't bother them. They know what I do, and they are understanding." She takes her responsibility seriously, carefully deciding what to post in order to strike the right tone. "I can't imagine doing anything else—I love it."

There are eight administrators that run the EDL's Facebook page, each responsible for finding and posting relevant articles, providing advice about upcoming demonstrations, deleting inappropriate comments, answering direct messages they receive, thanking supporters and tackling trolls. "We get a lot of them," one of the admins tells me. According to Hel Gower—Tommy Robinson's PA (although she'd be more accurately described as a "fixer")—one of the most time-consuming jobs the EDL Facebook admins have is getting rid of racist invective. This job is made more difficult by the fact that the EDL's Facebook page is also followed by a lot of anti-EDL users, people who masquerade as fans, but are only there to cause the group trouble. Each admin spends about an hour a day dealing with all this.

Because it's so important, the leadership keeps tight control over the admin and mod functions.* This means keeping a vise-

* Typically, an administrator is in charge of the entire page or group, while a moderator has specific powers to edit or delete other users' posts.

like grip on the passwords. In 2010, a member of a splinter group successfully convinced the admin of a local EDL branch Facebook page to give him their password. The newcomer swiftly changed the password, locked out the old admin, and hijacked the page. It took two weeks for Tommy Robinson to wrest back control, but he eventually managed to obtain the new password. I asked him how.

"A few lads went round there and got the password back," he says.

"How did they do that exactly?"

"We just made sure we got it back," he replies.

Paul began spending increasing amounts of time as the password holder in his group, sharing stories, and building up a virtual network of friends. It was as much social as it was political. There was a sense of solidarity and camaraderie that came with membership. "We were all against the same things, and we felt like a team making a difference," he says. But a virtual community can also become suffocating. The more time he spent online, the more extreme his views became. He became very concerned about Islamists, and the threat he thought they posed: "I learnt how sophisticated their tactics are, how they are trying slowly to steal our identity, take over our politics." It was also here, in these raucous and aggressive Facebook pages, that he first started to interact with Muslims directly. He found them every bit as angry as he was. Each interaction seemed to push him on, to increase the intensity and number of his attacks. And his adversaries were more than willing to fight back. "Scum! Subhuman scum," he fumes at me, recalling the "battles" he has had. These online tussles were an important part of Paul's daily routine—and consumed more and more of his time. How long did you spend on an average day on the internet? I ask. "It'd probably shock me if I worked it out." (He later estimates it to be 90 percent.) "It didn't leave much room for anything or anyone else," he says, confessing that during this time he became "a little bit of a sociophobe." He started speaking to his parents less and less, because it seemed "so mundane" compared to the conversations he was having online. As his online profile grew, so his real-world profile diminished.

Paul and I spent some time walking around his small town. There is very little to do there. Paul tells me he'd love to get into politics, in some way, and move to a bigger city, but with little employment experience, few qualifications and no money, he realizes that there's very little chance of either happening. He tells me that not long ago he walked past a group of EDL supporters. He didn't speak to them. Online he was becoming a respected member of the nationalist scene, with friends and supporters from all over the world. Offline he was nobody.

THE BATTLE FOR CYBERSPACE

In early 2012, Paul decided to strike out on his own. He found the clutch of traditional nationalist parties a bit staid and old-fashioned. Rather than settle for what was there—and persuaded by his powers of rhetoric—he started a new movement instead. He spent weeks learning how to make videos, and set up a personal blog, Twitter and Facebook accounts. Paul took quite some time making sure the imagery and visuals were just right. "I was trying to create a symbol that everyone could look to—a solid symbol." His experience on Facebook had persuaded him to adopt a secret, anonymous profile where he could be more honest without fear of reprisals.

Paul became increasingly embroiled in what is a running battle online between nationalists and antifascist opponents ("antifa"). Far-right groups and antifa used to clash on the street—they still do—but now the battle is mostly waged online. Antifa groups monitor every move the EDL and others like Paul make online, constantly watching key accounts, attempting to infiltrate their groups, and taking screen grabs or "screenies" of anything they consider controversial, offensive, or illegal, which they immediately publicize and often send to the police.

The longest standing of these groups is Exposing Racism and Intolerance Online, usually abbreviated to Expose. It's an online collective primarily based on Facebook and Twitter, with a dozen or so admins and perhaps a couple of hundred volunteers who help out occasionally. Their main activity consists of taking and saving screenshots of far-right communication and propaganda. Over the last four years Expose has amassed at least 10,000 of these screenshots, including some that first linked Anders Breivik to the EDL.

Antifa is full of a new type of citizen activist. Mikey Swales has been involved from the start. I contacted him via Facebook. "We're just an ordinary bunch of folk," he says, "mothers, fathers, sons, and daughters. We recognize racism, hatred, and bigotry when we see it and help, with other antifa groups, to show folk out there exactly who and what make up the EDL and all their splinter groups." Antifas spend just as much time as Paul online. One lone vigilante uses the Twitter handle "@Norsefired." He monitors EDL activity, and publishes around one hundred tweets a day, "challenging, exposing, and ridiculing extremist groups." Like Paul, he got involved by accident, when he caught some slack on Twitter for being part of an anti-cuts group and found out one of the attackers had an EDL link. And like Paul, he thinks he spends too long online: "My ear [is] getting bent from my other half," he tells me via email, "that my spare time should be more efficiently applied to more lucrative pursuits." @Norsefired thinks using a pseudonym allows him to confront his opponents more aggressively. Offline, he reckons, "it is unlikely I'd approach a group of EDL supporters. But my Norsefired persona can be quite no-nonsense, direct, cutting." One of his favorite tactics is to "occupy" EDL Twitter users' timelines—using several fake Twitter accounts to befriend as many of them as possible—then posting anti-EDL stories and news items simultaneously. One Expose member, Alex, explains to me that humor is a very important part of what they do. "Basically," he says, "I take the piss. I have a large archive of pictures and videos I've made to mock the right with." Their diverse methods can be quite effective. When it was alleged that the glamour model Katie Price was a supporter of the EDL, Alex managed to get in touch with her and persuade her to publicly deny it.

If you're antifa, infiltrating the "closed" groups—those that require permission or passwords to join—is the real prize. To do this, antifas often create fake accounts (or "sock puppets") and pretend to be sympathetic to the EDL cause. Sometimes one person controls dozens of different sock puppets, each with their own personality and affiliations. I spoke to one activist who has spent two years working up his—carefully liking certain pages, posting appropriate comments, building up a friendship network. Most forums and pages—whether EDL or antifa—are drowning in fake accounts. Tommy Robinson told me that almost every EDL group has been infiltrated, "by both far-right splinters and left-wing activists." "Do your people infiltrate as well?" I ask. He looks slightly coy. "Well, there might be some people that do that sort of thing, find out what they are saying about us, but I don't ask them to do it," he says.

In truth, both sides are at it. One Expose group recently outed a far-right sympathizer who had joined over 650 Facebook groups, including hundreds of left-wing and anti-fascist groups. She started by posting messages of support to gain trust among the antifa groups, and then kept a low profile and silently watched, in order to capture information about their tactics and targets. On the other side, infiltration has been an endless problem for the Casuals, a far-right organization with roots in football hooligan culture. Last year, a "Free the Brierfield 5" (jailed EDL supporters) group was set up by antifa as a trap, and a few enthusiastic Casuals joined it, sharing valuable information. "Some of us have learned nothing in three years of being stalked by these online weirdos," fumed Joe "Stabby" Marsh at other members on the Casuals' blog.

> Every time something happens that they know patriots will be angry about, they set up groups to trap people into saying things in anger that they hope they can get you nicked for. If you list where you work, or mention it in convos, they will screenshot you and they WILL ring and email your employer to try and cause you shit.

The sophistication of some of these groups is remarkable. The Cheerleaders are an unusual mix of Muslims, atheists, fashion models, and former soldiers drawn from all over the world: mostly female, and mostly highly skilled programmers. The closest thing to a leader in this leaderless collective is probably Charlie Flowers, a former punk-rock musician in his early forties who had some sympathy with the EDL at the beginning, but left as the group drifted in a more extreme direction. The Cheerleaders are committed to tackling any type of extremism online. A few dozen of them from all over the world frequently meet in a secret Facebook group to plan their actions. They have also been called "cyber-thugs" by enemies. It's an unfair claim: although they occasionally act as hired help for causes they agree with, their methods are legal, if a little devious. Charlie has managed to get a few websites shut down by placing a Digital Millennium Copyright Act notice on his own webpage, and then waiting for his adversaries to screenshot and use his material without permission—which he instantly reports. They can appeal, of course, but only if they sign a public affidavit with their real names and addresses, which lots of bloggers would rather not do. "A potent weapon, if used correctly," Charlie told me, chuckling. The strangest of all the ruses I witnessed was a Facebook page set up soon after the murder of Lee Rigby, by someone claiming to be part of antifa. It was called "Lee Rigby deserved it." The admin, who posted a picture of himself, declared: "I work for Hope not Hate [an antifa group] and the Community Party." He went on: "I Believe that Lee Rigby has become a far-right martyr and his death is being used as an excuse for violence and EDL buggery and thuggery. I want to lead the Communist revolution and take to the streets of Great Britain and declare this country the Soviet British Union." The real owner of this page was not antifa at all, but a radical right-winger who (I think) hoped to push EDL members into a more extreme view of the antifa groups. Although quite obviously a very poor and transparent effort, it seemed to work: the page exploded with fury for hours. A user named Dave threatened to "cave ur fucking skull in you bastard," while another named Kevin declared he was going to track the poster's home address: "Good luck cunt in surviving the week."

With all this trickery, working out who's who can be incredibly difficult. Fiyaz Mughal, the head of Tell Mama—a group set up to find and document anti-Muslim hate—now hires internet detectives who use open-source information to try to piece together real-world identities and their networks. Even then, he says, "we're only ever sixty percent sure." Trying to find out someone's real-world identity and linking it to their offline one—doxing them—is a common but very controversial tactic used by both sides, because it runs squarely against internet etiquette, and can also be extremely damaging for the person who's been discovered. SLATEDL and Expose—two of the key antifa groups—fell out over whether publishing home and work addresses and going after users in real life was acceptable. Mikey from Expose told me that doxing "is an absolute no-go area and it will never, ever be allowed in our group." Hel Gower, however, told me that someone from Expose had posted her personal information on their Facebook page, which they'd found on the Companies House register (I put that to Mikey, who told me they only post information that is already in the public domain, which would include the Companies House register).

The most infamous doxing site is RedWatch, a far-right

website that was set up in 2001. Its self-confessed aim is to find and identify traitorous British "lefties," posting the addresses, workplaces, children's names, and any other information they can obtain about those they believe are guilty of harassing and assaulting "British Nationalists and their families." It is infrequently updated, but retains a certain notoriety online. In 2003, two people who appeared on the website had their car firebombed. Paul and @Norsefired both fear being doxed, for slightly different reasons. Paul won't ever use his real name online, although he says he'd love to, because of the death threats he receives. @Norsefired worries that his name will end up on RedWatch. He insisted I didn't give anything away that might identify him. Doxers seem to know no limits, but there is often little the police can do unless a direct and specific threat against someone is made. People go to great lengths to dox others. In 2010, two hacking collectives, ZCompany Hacking Crew and TeaM PoisoN, managed to hack the EDL's Facebook account and take down their main page. The following year, TeaM PoisoN hacked the EDL again, and leaked details about the leadership of the group—phone numbers, email addresses, home addresses, and even the user names and passwords of the admins who run all the sites (including some rather funny passwords: Cameron, Winston1066, Anglosaxon1, and allah666, to name but a few).

While I was with Paul, he showed me how an antifa had attempted, and almost succeeded, to dox him. He was beginning to feel under siege. "I can feel myself becoming more radicalized by what these people are doing to me," he says. "I'm not a violent person. But I could happily, easily, see these scum suffer."

THE DENOUEMENT

Many nationalists feel completely disconnected, frustrated, and angered by traditional politics—and sometimes with good reason. Sitting with Paul in a run-down pub, the world of Westminster feels a very, very long way away. Queen Lareefer had never voted before becoming part of the EDL: "I feel ashamed that I took for granted the democracy that was given to me." Tommy Robinson left the EDL in late 2013 to try to pursue his ideas less violently. He was a football hooligan, and now he has plans to create his own political research group for the working classes. Whatever their beliefs, the internet, and social media have made politics accessible and appealing to countless people, and that has to be a good thing.

On the other hand, the same dynamic allows hundreds of small, often closed communities and individuals to surround themselves with information and people that corroborate their world view, and gives violent racists and xenophobes a platform on which to spread their message quickly and effectively. Creating our own realities is nothing new, but now it's easier than ever to become trapped in echo chambers of our own making. Nationalists and antifas both surround themselves with information that confirms what they already think. That can take people in a very dangerous direction. Breivik had convinced himself Norway was on the brink of destruction. Paul's echo chamber has led him to believe that whites are "beautiful, intelligent, artistic, creative, magnanimous," but now in a "tiny minority" as millions of migrants ("sneering, violent, drug dealers") are taking over England. In Paul's universe—lived through a screen—that's his reality. I remind him that Britain is 85 percent white, but he won't believe it.

Paul genuinely believes he is standing up for the country and its culture, facing down an existential threat from radical Islam. Antifa believe that fascists are on the march across the country, that everyone in the EDL is a closet racist and violent thug, and that they are facing down a possible resurgence of fascism in the country. The reality is far more nuanced, but in their own closed universes they are both right. In their personal echo chambers they've made demons and enemies of each other. Neither is as bad as the other thinks.

Throughout our months together, I tried to understand which of these forces—the echo chamber or the public sphere—exerted the stronger pull on Paul. Online he always seemed so violent and aggressive, although he was clearly quite proud that he had become a voice in the public debate. Although Paul told me that he considers Breivik "a hero," he also strongly denied that he would actually hurt anyone in spite of his extreme language. But I became worried about where this all might take him. There are a lot of people screaming hate online. Although only a tiny proportion will ever commit a violent act, it's almost impossible to tell who that might be. Yet whenever we met in person, I was a little assuaged. Paul's diatribes were usually prefixed with an apology. For him, the online and offline worlds were clearly very different realms.

But when Paul vanished suddenly, I started to worry. I feared his two worlds had collapsed: perhaps the police had tracked him down. Or worse. Two months later, I received an email from an unknown address. Paul hadn't gone anywhere—he just needed a break. "I was becoming too hate-filled, too paranoid, it was seeping into my blood, my bones," he tells me. He felt under too much pressure from all the trolling and abuse, and worried about the effects the attacks were having on him. He decided to kill off the digital Paul he'd created. "It was hard, because I ache to have a voice."

The last time I heard from Paul, he had created a new online persona, a woman, whose profile he was busily building, posting in the comment sections of political websites. Slowly, tentatively, but very deliberately, he was trying to drag a few more people towards his view of the world from behind a computer screen.

3

INTO GALT'S GULCH

We the Cypherpunks are dedicated to building anonymous systems.

—Eric Hughes, "A Cypherpunk's Manifesto" (1993)

A large abandoned Pizza Express in north London is an unusual place to start a revolution. But seventy of us have turned up. There are a number of speakers, but most of us are here to listen to a computer coder named Amir Taaki explain how the crypto-currency Bitcoin will change the world. We share the space with a dozen slightly confused-looking squatters who have recently taken up residence here. Cans of lager are being passed around, and there is a fug of cigarette smoke in the air, which gives the whole event a rebellious edge, especially for the nonsmoking sedentary audience members like me. There is a hush, as an unshaven man with short dark hair and a thin ponytail walks to the front of the room. Amir is in his mid-twenties, but is already considered to be one of the most gifted computer programmers around. In 2014, Forbes named him as one of the world's top thirty entrepreneurs under thirty. He is frequently offered lucrative jobs in the tech sector, but lives instead in what he calls a "techno-industrial colony" in Calafou, Spain. He's been working on Bitcoin software day and night for over four years now, and arguably knows more about this strange new currency than almost anyone else alive. He is here to tell us about his latest Bitcoin project—something he calls the "Dark Wallet."

The reason Amir and so many others like him are excited by Bitcoin is that it's a form of internet money with potentially far-reaching consequences. A Bitcoin is nothing more than a unique string of numbers. It has no independent value, and is not tied to any real-world currency. Its strength and value come from the fact that people believe in it and use it. Anyone can download a Bitcoin wallet on to their computer, buy Bitcoins with traditional currency from a currency exchange, and use them to buy or sell a growing number of products or services as easily as sending an email. Transactions are secure, fast, and free, with no central authority controlling value or supply, and no middlemen taking a slice. You don't even have to give your real name to start up an account. No one person or group is in charge of Bitcoin: everyone is.

Bitcoin was introduced to the world in 2009 via a public post on an exclusive emailing list for cryptographers. It quickly developed a following, and soon became the currency of choice for the online drug market Silk Road. A growing number of people started to exchange Bitcoin for dollars, which pushed its exchange rate from under $0.001 in October 2009 to $100 in April 2013. In

October that year, a U.S. Federal Reserve spokesman hinted that Bitcoin might one day become a "viable currency," and the following month the value of a single Bitcoin jumped to over $1,000. Millions of dollars' worth of Bitcoin are now traded every day (as of writing, a single Bitcoin is worth around $300). More and more commercial outlets are now accepting the currency. In some parts of the world you can live almost entirely on Bitcoin.

Bitcoin's dramatic rise to prominence resulted in an explosion of investment, exchange companies, and even ATMs (my favorite is a company called BitPesa, which helps people send overseas remittances to Africa for almost no cost). Many members of the Bitcoin community have entered into complex negotiations with governments and regulators about how to make the new digital currency work alongside the traditional. The Bitcoin Foundation, a semi-official body that represents the currency, was established in 2012 in order to standardize the core development required to keep the system working securely and effectively. Although no one is really in charge of Bitcoin, the Foundation is probably the closest thing there is to a governing body. In 2013, the Bitcoin Foundation's annual conference was called "The Future of Payment," a title that reflects the view of many of its users: that Bitcoin can be part of the system. But not all of them.

Amir begins his talk about the Dark Wallet by describing some of the technical challenges he's faced, but soon drifts into a polemic. "Bitcoins aren't a fucking payment innovation," Amir shouts. "Bitcoins are a political project."

"Maybe we should work with governments?" one audience member suggests. "Wouldn't it help to extend Bitcoin's reach?"

"No!" Amir replies. "The government is just one big bunch of gangsters! You can't placate gangsters! Right now, it's us who have the initiative. And we're not going to give it back."

For people like Amir, Bitcoins are the front line in a bigger battle over the right to anonymity and freedom online. Amir believes you should be free to be whoever, say whatever and do whatever you want online without censorship or surveillance—and that such freedoms will lead to political revolutions. He is a cypherpunk.

THE MAILING LIST

One day in late 1992, retired businessman Tim May, mathematician Eric Hughes and computer scientist John Gilmore—the creator of *alt.*—invited twenty of their favorite programmers and cryptographers to Hughes's house in Oakland, California. After taking a degree in physics at the University of California at Santa Barbara, May went to work for Intel in 1974, where he made a brilliant breakthrough in redesigning Intel's computer memory chips. He retired at the age of thirty-four and dedicated himself to reading: computing, cryptography, physics, mathematics—and to politics. Gilmore was Sun Microsystems' fifth employee; like May, he retired young to pursue political ideas. Hughes, a brilliant mathematician from the University of California, Berkeley, had spent time working in the Netherlands with David Chaum, perhaps the world's best-known cryptographer at that point. May, Hughes, and Gilmore were natural bedfellows. All were radical libertarians and early adopters of computer technology, sharing an interest in the effects it would have on politics and society. But while many West Coast liberals were toasting the dawn of a new and liberating electronic age, Hughes, May, and Gilmore spotted that networked computing might just as likely herald a golden age of state spying and control. They all believed that the great political issue of the day was whether governments of the world would use the internet to strangle individual freedom and privacy through digital surveillance, or whether autonomous individuals would undermine and even destroy the state through the subversive tools digital computing also promised.

At their first meeting, May set out his vision to the excited group of rebellious, ponytailed twenty- and thirtysomethings. If the government can't monitor you, he argued, it can't control you. Fortunately, said May, thanks to modern computing, individual liberty can be assured by something more reliable than man-made laws: the unflinching rules of math and physics, existing on software that couldn't be deleted. "Politics has never given anyone lasting freedom, and it never will," he wrote in 1993. But computer systems could. What was needed, May argued, was new software that could help ordinary people evade government surveillance. The group was set up to find out how.

Soon the group began to meet every month in the office of Cygnus Solutions, a business that Gilmore had recently set up. At one of the first meetings in 1992, one member—Jude Milhon, who wrote articles for *Mondo 2000* under the alias St Jude—described the growing movement as "the cypherpunks," a play on the cyberpunk genre of fiction made popular by sci-fi writers such as William Gibson. The name stuck. "It was a bit of a marketing ploy, to be honest," May told me over the phone from his home in California. "A bit like Anonymous wearing the Guy Fawkes masks."

The group began to grow. Eric Hughes decided to set up an

email list to reach out to other interested parties beyond the Bay Area. The list was hosted by the server that ran Gilmore's personal website, toad.com. The first post on the list, even before the introduction from Hughes, was a repost of a 1987 speech given by mathematician Chuck Hammill called "From Crossbows to Cryptography: Thwarting the State via Technology." It set the tone perfectly for what would follow: "For a fraction of the investment in time, money, and effort I might expend in trying to convince the state to abolish wiretapping and all forms of censorship," wrote Hammill, "I can teach every libertarian who's interested how to use cryptography to abolish them unilaterally." The list quickly grew to include hundreds of subscribers who were soon posting every day: exchanging ideas, discussing developments, proposing and testing cyphers. This remarkable email list predicted, developed, or invented almost every technique now employed by computer users to avoid government surveillance. Tim May proposed, among other things, secure crypto-currencies, a tool enabling people to browse the web anonymously, an unregulated marketplace—which he called "BlackNet"—where anything could be bought or sold without being tracked, and a prototype anonymous whistleblowing system.

The cypherpunks were troublemakers: controversial, radical, unrelenting, but also practical. They made things. Someone would write a piece of software, post it to the list, and others would test it and improve it. When Hughes put forward a program for anonymous remailers—a way to email people without being traced—another influential poster to the list, Hal Finney, worked to correct a flaw he'd spotted in it, then posted his improved version. Among the cypherpunks, writes *Forbes* journalist

Andy Greenberg in his history of whistleblowing, creativity was more admired than theorizing. It was Hughes who coined the expression that would define them: "cypherpunks write code."

Above all, the code they wanted to write was encryption. Encryption is the art and science of keeping things secret from people you don't want to know them, while revealing them to those you do. From the time of the Roman Empire until the 1970s, encryption was based on a "single key" model, with the same code both locking and unlocking the message. Modern computing made encryption far more powerful, but the underlying principle was the same: if you wanted to communicate secretly with someone, you still had to get the code to them—which presented the same problem you started with. Two MIT mathematicians called Whitfield Diffie and Martin Hellman solved this in 1976 with a system they called "public key encryption." Each user is given his own personal cypher system comprised of two "keys," which are different but mathematically related to each other through their relationship to a shared prime number. The mathematics behind it is complicated, but the idea is simple. It means you can share your "public" key with everyone, and they can use it to encrypt a message into a meaningless jumble that can be decrypted only with your secret "private" key. Public key encryption transformed the potential uses of encryption, because suddenly people were able to send encrypted messages to each other without having to also exchange a code, and indeed without even having to ever meet at all. Up until the early nineties, powerful encryption was the sole preserve of governments. The United States had even classed powerful encryption as a "munition" in 1976 and made its export illegal without a license.

As more people ventured into cyberspace, the U.S. government began to take a greater interest in what they were doing there. In 1990, the FBI launched an over-the-top crackdown on computer hackers, known as Operation Sundevil. This was swiftly followed, in early 1991, by a proposed piece of U.S. Senate legislation that would force electronic communications service providers to hand over people's personal data. (The key clause, S.266, was pushed by the then chairman of the U.S. Senate Judiciary Committee, Senator Joe Biden.) Worse still, in 1993 the U.S. government announced the "Clipper Chip": industry standard encryption for the internet, to which the National Security Agency would hold all the keys.

Many early adopters of the net considered this to be an attempt by the U.S. government to control cyberspace, which until that point had operated largely outside state control. Phil Zimmermann, an antinuclear activist and computer programmer, was worried that digital technologies appeared to be eroding citizens' privacy, rather than liberating them. For years, Zimmermann had dreamed of creating an encryption system for the masses based on public key encryption that would allow political activists to communicate free from the government's prying eyes. However, juggling a freelance job and two children, he had never found the time to realize it. On learning of Biden's S.266 clause, he feverishly set out to complete the project, almost losing his house in the process. When Zimmermann finished his software in 1991, he published it all online—on a Usenet group, of course—free for anyone who wanted to use it. He called it "Pretty Good Privacy," or PGP for short, and within weeks it had been downloaded and shared by thousands of people around the world. "Before PGP, there was no way for two ordinary people to communicate over long distances without the risk of interception," said Zimmermann in a later interview. "Not by phone, not by FedEx, not by fax." It remains the most widely used form of email encryption to this day.

The U.S. government, needless to say, wasn't happy. They believed too many people using strong cryptography like PGP would make life a lot harder for the security services. The British government was also watching nervously. Sir David Omand, who was working for the British intelligence agency GCHQ at this time, recalls the period well. "We were very worried about the spread and adoption of powerful encryption like PGP," he said. The British government even briefly considered following France in legislating to control encryption. In the end, they decided against: once Zimmermann had released the source code online, it was going to be almost impossible to try to remove it from the public domain. Besides, it was increasingly obvious that encryption technology was vital for the health of the rapidly expanding internet, especially for online trade and commerce. A more secure internet would be trusted by more people. The U.S. government decided on a different course. Zimmermann, having released his PGP source code on the internet, was considered by the U.S. government to have exported munitions. The United States Customs Service launched a criminal investigation, seeking to prosecute Zimmermann under the Arms Export Controls Act.

This battle over encryption became known as the Crypto-Wars, fought between those who believed citizens should have the right to possess strong cryptography, and the government who did not. For May, Gilmore, and Hughes, making sure crypto was available to all was a means to an end. The cypherpunks hoped and believed their endeavors would eventually bring about an economic, political, and social revolution. Their list fizzed with political radicalism. In 1994, May published *Cyphernomicon*, his manifesto of the cypherpunk world view, on the mailing list. In it, he explained that "many of us are explicitly antidemocratic and hope to use encryption to undermine the so-called democratic governments of the world." On the whole, the cypherpunks were rugged libertarians who believed that far too many decisions that affected the liberty of the individual were determined by a popular vote of democratic governments. The cypherpunks were advised to read *1984*, the cult science-fiction novels *The Shockwave Rider* and *True Names*, David Chaum's paper "Security without Identification: Transaction Systems to Make Big Brother Obsolete" and perhaps most importantly, *Atlas Shrugged*. In Ayn Rand's magnum opus, the most productive citizens of a dystopian American society refuse to pay taxes and disappear to "Galt's Gulch," a secluded community whose inhabitants are free to pursue greatness. May hoped to see similar "virtual regions" where individuals could make consensual economic arrangements among themselves with no state at all.

The mailing list became the favorite watering hole for hundreds of talented computer programmers and hackers from all over the world, many of whom would use the list to learn about crypto before setting out to pursue May's vision in their own way. One of them was a programmer named "Proff," who joined the cypherpunk mailing list in late 1993 or early 1994. He immediately got sucked into the raucous and aggressive exchanges that characterized the cypherpunks: insulting newcomers, ruthlessly criticizing perceived shortcomings in others' technical knowledge, and plotting the downfall of governments. When Esther Dyson, head of the Electronic Frontier Foundation (EFF)—a civil liberties group committed to free internet expression and privacy, cofounded by Gilmore—argued on the list that some limits to anonymity might be acceptable if there were very strict laws respecting privacy, Proff shot back: "It is clear that the personal beliefs of those involved in EFF are those of compromise, present-day politics, and a general lack of moral fiber." Proff even speculated that Dyson worked for the CIA. Dyson replied, "For the record, I am not a tool of the CIA nor have they pressured me, but there's no reason for you to believe me."*

"Proff," it transpired, was a gifted young Australian programmer called Julian Assange. Although Assange was a libertarian, he did not share May's unashamed elitism: in the *Cyphernomicon* May spoke disparagingly of "nonproductive" citizens, "inner-city breeders" and, most notoriously, the "clueless 95 percent." In one of his last posts on the list, Assange wrote (likely in rebuttal to May) that "the 95 percent of the population which compromise the flock have never been my target and neither should they be yours. It's the 2.5 percent at either end of the normal that I have in my sights." (When I asked May if he thought Assange was a "true" cypherpunk, he replied, "Yes, absolutely. I count him as one of us. He did things, he set things up, and he built things.")

May's dislike of government appears to have been an intellectual discovery, the product mainly of voracious reading. For Assange, it was more emotional. In 1991, he had been arrested for hacking into the Australian telecom company Nortel, under the

* Despite their heated exchanges, the two would later become friends.

pseudonym Mendax. Although he avoided prison, the threat of criminal prosecution had lingered over him for two years before he pleaded guilty to twenty-five hacking charges in 1994. The experience, he later wrote, allowed him "to see through that veneer the educated swear to disbelieve in but still slavishly follow with their hearts!"

Assange saw that crypto could be used for offense as well as defense. He believed that the anonymity crypto could provide would facilitate and encourage whistleblowers to expose state secrets. For Assange, crypto could pry governments open, making them more transparent—"to see through that veneer"—and more accountable, and hopefully pull down a few in the process. His inspiration came from another cypherpunk from the mailing list named John Young, who in 1996 founded the website cryptome.org as a place to publish leaked documents—especially any confidential government records and reports. Assange had contacted Young in 2006, saying, "You knew me under another name from the cypherpunk days." He told Young of his plan to create a new organization, which he called WikiLeaks, which he believed would change the world: "New technology and cryptographic ideas permit us to not only encourage document leaking, but to facilitate it directly on a mass scale. We intend to place a new star in the political firmament of man."

For almost a decade, the cypherpunk mailing list was the center of the crypto world. Hundreds of people used it to propose and learn ciphers, evade detection, discuss radical politics. It was finally discontinued in 2001 when John Gilmore booted it from his host, toad.com, for reasons not entirely clear—Gilmore claimed it had "degenerated." But it had a remarkable track record: anonymous remailers were everywhere, an anonymous browser that allowed users to browse the web without anyone being able to track them was in development, the whistleblowing site Cryptome was becoming a thorn in the side of intelligence agencies. Better still, the U.S. government had dropped its investigation into Phil Zimmermann, and PGP was being used all over the world.

One thing was still missing. Although the cypherpunks had tried to build a system of anonymous digital payment, they had never quite managed it. After Gilmore shut down the original mailing list, others sprang up in its place, with several dedicated to improving crypto. The most notable was the cryptography mailing list hosted by Perry Metzger, where many of the original cypherpunks migrated. But it also attracted a new generation who were just as keen to post papers and ideas about how to evade government surveillance and improve individual privacy online. In early 2008 a mysterious contributor to the cryptography mailing list called Satoshi Nakamoto posted a message that would change everything.

TO CALAFOU

Six weeks after Amir's talk, I find myself walking down a dusty hill and over a concrete bridge towards an enormous nineteenth-century textile factory complex. The words "Calafou: *còlonia ecoindustrial postcapitalista*" are painted in large black and green letters on a wall outside. It's midafternoon. I approach a bearded, long-haired man loitering by the entrance, and ask for Amir. "He'll be in the hackers' space," he says. "Or asleep." I wander in.

Calafou is an experiment in cooperative living. It is currently run and managed by its thirty or so permanent residents, in partnership with an organization called the "Catalan Integral Cooperative" (CIC). CIC's vision is to find new ways of living sustainably, ethically, and communally outside the capitalist system, based on the principles of economic and political self-determination.

Everything about Calafou is big. The plot must be 200 acres, although I can't really tell because it's so busy with buildings. There are around thirty or so flats, each containing four small rooms, and over 10,000 square meters of former industrial floor space, including a communal dining area and an old abandoned church, which served the spiritual needs of the factory workers who used to live here. The place looks to be in a near-permanent state of creative destruction, cluttered with motorbike engines, half-built bicycles, a row of plasterboard, empty beer bottles, a tractor tire on its side, a pile of bricks, two 3D printers. I eventually find the hackers' space towards the back of the complex. It is reached through a large, roofless hall, and up a couple of flights of concrete stairs. It's about the size of a tennis court, packed with old computers, boxes full of modems, wires, cables, and telephones (I later learn every computer they have is recycled or secondhand). A couple of worn sofas line the far wall and a large table in the middle houses more computers, food, and a landline telephone. A huge spray-painting of Captain Crunch, the infamous telephone hacker from the 1970s, and Alan Turing, the genius British cryptographer, leaves little doubt about the group's loyalties.

There are a few people coding—two young men in one corner, and a slightly older man in a hoodie sitting in front of three computer screens, smoking a cigarette. He's deep in concentration.

This must be Pablo, Amir's chief collaborator. Pablo is responsible for the "front end" of the Dark Wallet, the bit you see on your computer. I walk in. No one looks up from their computer. I introduce myself to Pablo, and ask if he has time to talk. He doesn't, he says, because he's grappling with a programming problem, but he might soon. I sit down on one of the sofas. This is how most computer programmers and hackers work. Advanced coding is a creative endeavor, Amir had told me back in London. When you're on a roll, you keep going. Pablo was obviously on a roll.*

Eventually he stops typing, expertly rolls another cigarette, and joins me on the sofa. We begin talking about the factory. Pablo is a full-time resident at Calafou. He tells me it's an exciting time at CIC because the residents are currently in negotiations to buy the entire factory complex, with each person paying 25,000 euros for a flat. For now, it's all rented. Just over 100 euros will get you a room and working space for a month. Throw in the communal cooking system, and you can get by on very little, and be free to develop your own projects (when you're not putting in some free labor for the community). Dark Wallet is one of dozens of projects at Calafou, Pablo says. Just before I arrived there had been a session on 3D printing. In the room next door, there is a scientific experiment to grow a strand of amoeba that can store energy. The long-term plan is to create organic computers. Other residents are creating compost toilets, manufacturing solar panels, selling clay ovens, and building open-source telecommunications. All the

* Someone told me later he sometimes works forty-eight hours straight, before sleeping for a day to recover. It turns out, when Pablo finally does talk, that he'd been in the middle of receiving the first successful Bitcoin transaction using a "stealth address," one that cannot be traced.

apartments are now full, but there are always extra people couch surfing, especially if there is a public event on, which is often.

Calafou is more than a living space, says Pablo. It's also a philosophy, inspired and partly funded by a man called Enric Duran. "He's an amazing man," says Pablo excitedly. Indeed he is. In late 2008, Duran—dubbed the Robin Hood of the banks—circulated 200,000 copies of a free newspaper called *Crisis* to explain how he had spent the previous two years fooling thirty-nine banks into lending him nearly half a million euros. He paid back early loans to ensure a good credit rating, borrowed more, stopped paying, and gave it all away to social activists (including, I was told, to Calafou) and to pay for future editions of *Crisis*. In 2009, Duran began promoting the CIC as a practical example of the ideals detailed in his second newspaper: *We Can! Live Without Capitalism*. He was arrested in 2009 on charges brought against him by six of the banks, and spent two months in prison before being released on bail. When, in 2011, the state prosecutor requested an eight-year prison sentence against Duran, he went into hiding.

After an hour or so, Amir rolls lazily into the room with two friends from the broadly anticapitalist protest movement Occupy London, who are visiting. He doesn't notice me, and doesn't really register Pablo either. "Amir," shouts Pablo, "I've received the first Bitcoin transaction from the stealth address!" Amir stares intensely at Pablo's screens for a moment, nodding slowly as his eyes dart about hurriedly. He seems fairly unmoved. "Cool," he says.

Amir was born in London to an Iranian dad and a Scottish-English mother, but was raised in nearby Kent. He taught himself computer programming at school, and quickly found himself in trouble after shutting down the school's CCTV network. He excelled at math, and eventually went on to study it at university—a course he started and abandoned three times. He became a squatter, and met Pablo, with whom he then spent five years working on an open-source computer game. Shortly before it was due to be released, the project collapsed. "Politics and people got in the way," Amir explains. "I suddenly found myself with no money and no education. I felt that I'd just wasted five years." Although he and Pablo got on well, the experience of working in a large team was not a success: "The worst thing in your life," he adds, "is listening to other people."

He then spent an increasing amount of time online, making money as a professional poker player. For two years, he played hundreds of poker hands a day, multiple hands at once. He didn't earn a fortune—but enough to get by. This was the unlikely venue of his political education too. On "Black Friday," in April 2011, the founders of the three largest online poker companies in the United States were indicted in a criminal investigation and the FBI seized the websites. (In 2012, the U.S. government dismissed all civil complaints against PokerStars and Full Tilt Poker.) Thousands of players—including many of Amir's online poker friends—lost their money somewhere in cyberspace. Amir started experimenting with his own peer-to-peer poker site to cut out the online poker companies (and the stake they charge for each hand) but he couldn't find a decent, secure, payment system. Then in 2011 he stumbled across Bitcoin. He started working on a number of Bitcoin-related projects, and even founded and ran the UK's first Bitcoin exchange, called "Britcoin," which allowed people to exchange Bitcoins directly into pounds sterling, rather than via dollars. Digging around the Bitcoin protocols, he noticed it wasn't quite as secure and anonymous as everyone thought. It was a brilliant invention, of course, but with a few additions could be made even more subversive. That's when he came up with the idea of Dark Wallet. He moved to Calafou, brought in Pablo alongside Cody Wilson—the American crypto-anarchist who created the first 3D printed gun—and together they raised $50,000 in a month via the crowdfunding site Indiegogo.

Although Amir's technical know-how and experience are admired, his ideals and motivations have put him on the fringes of what has become an increasingly respectable Bitcoin community. Dark Wallet has pitted itself directly against organizations seeking to capitalize and control Bitcoin and its market. "Many prominent Bitcoin developers are actively in collusion with members of law enforcement and seeking approval from government legislators," reads the Dark Wallet blurb. "We believe this is not in Bitcoin users' self-interest, and instead serves wealthy business interests that make up the self-titled Bitcoin Foundation." In a 2014 interview with *Newsweek*, the chief Bitcoin Foundation scientist, Gavin Andresen, said that he thinks of Bitcoin as "a just-plain-better, more efficient, less-subject-to-political-whims money. Not as an all-powerful black-market tool that will be used by anarchists to overthrow the System." Some within the broader Bitcoin community worry that Amir's radical politics will stop the currency from being taken seriously. "This fuckwit Taaki takes the absolute cake . . ." wrote one on the popular Bitcointalk forum. "It is upon us as a community to cut them loose!" I emailed Mike Hearn, one of the chief programmers for the Bitcoin Foundation, who told me that, although he doesn't mind if government's power to control people through the banks is diminished, "I consider Bitcoin principally a technical project. I think people [like Amir] are going to be disappointed when it turns out that bankless money doesn't actually create anarchy."

Amir doesn't care about this sort of talk. A tool to overthrow the System is precisely how he sees it. "People at the Foundation are trying to censor Bitcoin," he tells me. Both he and Cody Wilson have been on record stating they hope Dark Wallet will be used to buy drugs more securely, and that any negotiation with governments betrays Bitcoin's vision. He fears its radical libertarian potential is being diluted. The Bitcoin Foundation, explains Amir, introduces censorship mechanisms or greater centralization (for example to mining) to help Bitcoin work more effectively and securely as a payment mechanism. But, says Amir, this also reduces its ability as an open-source technology that allows people to transact among themselves. He explains, "The Bitcoin Foundation says, 'Oh, we need to make it better for the consumers.' No we don't! What people forget is that Satoshi was political."

SATOSHI

Tim May and the cypherpunks hadn't invented digital crypto-currencies, but they'd seen what they might do. The honor goes to a cryptographer called David Chaum. Although he never attended a meeting, his work on anonymous payment systems was an inspiration for many cypherpunks, including May. The basic principle of a crypto-currency is that each unit of the currency is a string of unique numbers that users can send one another online. But strings of numbers can be easily copied and spent several times over, which makes them valueless. Chaum solved this problem by creating a single centralized ledger, which kept a record of each person's transaction to verify that each unit of currency wasn't in two places at once. He even set up a company in 1990 called DigiCash to realize these plans. But the idea of having only one central system verifying the whole network made it seem too unreliable to many. DigiCash never quite took off.

Satoshi's post on the cryptography mailing list proposed a new kind of digital crypto-currency, which he claimed solved this problem by creating a distributed system of verification. He called it Bitcoin. "He got a skeptical reception at first," recalls Hal Finney, the veteran cypherpunk who'd seen several proposals come and go. But Finney noticed Satoshi had included something he'd not really seen before, something called a blockchain.

A quantity of Bitcoin is stored at a Bitcoin address, the key to which is a unique string of letters and numbers that can be kept on a website, desktop, mobile phone, or even a piece of paper. Every time someone sends a Bitcoin as payment, a record of the transaction is stored in something called the blockchain. Transactions are collected into blocks, with each block representing about 10 minutes' worth of transactions. The blocks are ordered chronologically, and each includes a digital signature (a "hash") of the previous block, which administers the ordering and guarantees that a new block can join the chain only if it starts from where the preceding one finishes. A copy of the blockchain record—a record of every single transaction ever made—is maintained by everyone who has installed the Bitcoin software. To ensure everything is running as it should, the blockchains are constantly verified by the computers of everyone else using the software. The upshot of all this is that, at any point, the system knows exactly how many Bitcoins I have in my wallet, so they cannot be copied or spent twice. For the first time, ownership can be transferred, but never duplicated—and all without the assistance of a centrally controlled ledger. It is genius.

After Satoshi and Finney conducted the first ever transaction (and ironed out a few teething problems) Satoshi made it an open-source project, inviting others to help develop the code and the concept. More and more users joined the mailing list, and began to transfer Bitcoins to each other, always half expecting the system to break. But it never did.

The reason Bitcoin is so beloved by libertarians is because it takes control of the money supply away from the state. Satoshi distrusted the global banking system, and saw his cryptocurrency as a way to undermine it. He hated that bankers and governments held the key to the money supply and could manipulate it to their own ends. He even added an out-of-place line of text into the "genesis block" (the very first bit of the blockchain—his transactions with Finney), which read: "The Times 03/Jan/2009 Chancellor on brink of second bailout for banks."

To keep governments and central banks out of it, Satoshi placed a cap on the total number of Bitcoins that could ever be produced: 21 million. Although Bitcoins can be bought and sold with real-world currencies, new Bitcoins are not minted by any central authority. Instead anyone who dedicates their computing power to verifying the transactions in the blockchain competes to earn a very small amount of new Bitcoins each time they do so (this is called "mining"). As more Bitcoins are created (approximately 13 million have been created so far), the remaining Bitcoins require more computing power to mine.* The last Bitcoin is expected to be mined in around 2140. But it did not stop there. Satoshi designed it to be a peer-to-peer, encrypted, and quasi-anonymous system, which makes linking a Bitcoin transaction to a real-world person very difficult, thereby making collecting taxes and monitoring users extremely awkward. Even though the blockchain records the transactions, it doesn't record who is behind them.

These features are precisely what Satoshi had in mind all along. While many of his posts to the cryptography mailing list discussed the technicalities of the new currency, he also made his loyalties clear. In his early posts, Satoshi wrote on the list to Finney that Bitcoin was "very attractive to the libertarian viewpoint if we can explain it properly." "You will not find a solution to political problems in cryptography," wrote one poster in response. "Yes," replied Satoshi, "but we can win a major battle in the arms race and gain a new territory of freedom for several years."

Satoshi typed in his last post on the list in late 2010, and, like a true cypherpunk, promptly disappeared. Amir was right. At its core, Bitcoin is a political project. But it is also an open-source project, and for many, like the Bitcoin Foundation, its future is as a payment mechanism. For Amir, this is diluting the original cypherpunk vision. That's why he's building the Dark Wallet.

DARK WALLET

Happily ensconced in "Hackafou," Amir describes the aims of his project to me. Ultimately, it's about trying to make Bitcoins more

* Bitcoins can be divided into eight decimal places. The smallest non-divisible unit is known as a "Satoshi."

anonymous and more trustworthy. The Dark Wallet will include a number of new features, which, if implemented correctly, will certainly cause more of a headache for "The System." One of the key innovations is called "multi-signature," where a Bitcoin payment can be released only if two of three parties sign it off. Another is called "trustless mixing," a way of making Bitcoin payments harder to trace. It's based on a project called CoinJoin, which jumbles up transactions that are happening at the same time, and then reroutes them to the final destination. Everyone ends up with the right amount but no one knows who was sending what to whom. The third key innovation is called the "stealth address." Dark Wallet generates a fake Bitcoin address as the recipient's, meaning that it is a little harder to link a real person to their wallet. While not making Bitcoin transactions perfectly anonymous, this is a significant step forward. Amir anticipates that a lot of people will want to take advantage of the extra layers of security it offers.

As a computer programmer, Amir is exceptionally precise and exacting. But when I try to press him on the specifics of politics, it feels to me like he struggles to turn his frustration into a clear and coherent set of ideas. Whenever he speaks about Bitcoin and its potential, the conversation often quickly switches to angry polemic: the surveillance state, corrupt governments, greedy corporations, oppression, environmental damage. But, he explains, there is a simple and powerful idea that connects it all: decentralization. Technology, he says, empowers individuals. "I am for the human spirit, and against power," he says. He sees his role is to build the tools that other people can use to increase their spheres of freedom. (And it's true that Amir is deeply committed to sharing rather than profiting from the technology he produces. This is difficult not to admire.)

Like Tim May, he sees salvation in numbers, not man-made laws: "Bitcoin is a currency based on mathematics," he says. "The purest kind. It creates the purest market, peer to peer with no corrupt or controlling third parties." In that sense, he sees the Dark Wallet as one more strike against the inefficient and overly powerful governments in the world. "A bunch of gangster running a sham democracy," he says. Bitcoin cuts out the friction, the inefficiencies, that get in the way. (And as he will later explain, its potential exists in far more than money.) On this broad point, there are many within the Bitcoin community who agree with him.

It feels to me a little utopian: too much faith in the certainty of math and physics to resolve society's problems, with insufficient thought about precisely how to get there. Amir sees it differently. "I spend a lot of time in communities, observing the problems they have now. I look at the tools available to me, and see how to create solutions. There's no utopianism in this process, as it is an iterative one." But, I suggest, don't governments serve some useful purposes, even if they are imperfect? What about collective healthcare, education, help for those at the bottom?

Amir suddenly stops. "Do you want to play a computer game?" he asks. He loads up something called *Mirror's Edge*. The story is set in a near-future society in which a dictatorial state keeps the peace through a toxic mix of surveillance and sterile hyperconsumerism. The docile population prefers peace to freedom, except for a handful of rebels who rely on "runners" to deliver messages to the underground resistance. As a runner, your job is to scamper across the tops of building, scurry down back alleys, and disappear into the shadows, evading the state police. "I love games," Amir says. "They're how children learn about politics." He plays with his face impossibly close to the screen, head slightly cocked, half jumping out of his chair every time his online persona does. "Training," he says, chuckling. As he ducks and weaves he continues the thread we'd left off before he started the game: "It's true—people are going to suffer. Yes, that's sad. But that's just the way it is."

CYPHERPUNK GOES MAINSTREAM

Bitcoin is the means to an end for cypherpunks like Amir, just as it had been for Tim May. That end amounts to free forms of communication and transactions between individuals that cannot be censored or monitored. "Currencies are just the beginning," Amir tells me. "The real genius of blockchain is that it is going to help us create a decentralized net that no one can censor. This is much bigger than just Bitcoin. We're going to transform the entire internet."

"What do you mean?" I ask.

"Well, at the moment your Facebook data isn't really controlled by you: it's hosted on Mark Zuckerberg's servers. Facebook administrators can do anything they like with it, because they own the servers, and so they own your data. It's not really free, because it's centralized. A social media platform built using blockchain would be different. Your posts would become part of the public blockchain record, and every user of the platform would have their own copy. Everything could be done anonymously, and censorship would be close to impossible. No one can shut it down, because no one owns it."

There are several new projects under way that are trying to do this. One is a social media platform called Twister. Miguel Freitas is Twister's chief developer. Miguel worked for several straight months—also unpaid, just as Zimmermann did when working on PGP—to convert the blockchain model into a social media platform after the British Prime Minister, David Cameron, admitted his government considered shutting down Twitter during the 2011 London riots. "I tried searching for peer-to-peer microblogging alternatives, but I couldn't find any," he told me. "The internet alone won't help information flow if all the power is in the hands of Facebook and friends."

Others are at work on an even more ambitious project: to use the Bitcoin block chain system to transform the entire internet. The Ethereum project received around $12 million of crowd funded support when it was founded a couple of years ago, by a twenty-year-old Russian-Canadian programming wizard called Vitalik Buterin. That's been enough to hire forty of the smartest geeks you'll ever meet, and house them in comfort in Amsterdam, Berlin, and London. "Welcome to web 3.0," says Vinay Gupta, a hacker-cum-poverty-activist who's part of the Ethereum team, as I arrive at their London offices one evening in early 2015. Web 1.0 was all static websites, like bbc.com. Web 2.0 was interactive social media platforms like Facebook. This third iteration is about peer-to-peer networks backed by powerful encryption. It sounds a little dry, but Ethereum has London's tech crowd purring. Last year, Ethereum won the 2014 World Technology award for IT software. The mighty IBM has already used it to build a washing machine that orders its own soap. But these people aren't in it for the money: Ethereum is an open-source project, and all its employees will slink off when the project is complete. They're doing it because they want to transform the internet—and by extension, society.

Ethereum does two things, wrapped up into one piece of software. First, it's what Vinay calls "deep infrastructure." Essentially, Ethereum is building a new web out of the spare computing power and hard-drive space of millions of computers that its owners connect to the network. Your computer is like your brain—you only ever use a small amount of it. All that spare power and space can then be used by others to host software, apps, websites, that others can access. Because it runs with strong encryption and the network is "distributed" across all those computers, it's more or less impossible for anyone to censor or control what's on it. Second, it allows people to create immutable, public transaction records. (Bear with me on this: it's very important). Borrowing the idea from the blockchain, it will allow information to be recorded on a public database in a chronological way that prevents tampering, fraud, or deletion. So in sum: it's a new anonymous yet uncensored internet, and a new way of controlling and storing information.

Twister and Ethereum are only a couple of many next-generation systems to guarantee free expression, and privacy designed for the mass market rather than the specialist: each user-friendly, cheap, and efficient. Jitsi is a free, secure, open-source voice, video-conferencing, and instant-messaging application which started as a student project at the University of Strasbourg. Jabber, another instant-messaging service, is encrypted with industry-standard Secure Sockets Layer, run by volunteers and physically hosted in a secure data center. Phil Zimmermann is currently working on a project called Darkmail, an automatically end-to-end encrypted email service.

Today there are hundreds of people like Amir, Vinay, and Miguel working on ingenious ways of keeping online secrets or preventing censorship, often in their own time, and frequently crowdfunded by users sympathetic to the cause. One is Smári McCarthy. Smári is unashamedly geeky: a computer whiz and founding member of the radical Icelandic Pirate Party. He used to work with Julian Assange in the early days of WikiLeaks. Smári isn't really a cypherpunk—he resists any association with Ayn Rand's philosophy—but he does believe that privacy online is a fundamental right, and worries about state surveillance of the net. He also believes that crypto is a key part of a political project. He wants you to encrypt all your emails with PGP, even (or especially) those you send to friends and family members. The reason, he explains, is to provide "cover traffic" for those who do need to keep things secret. If everyone is using it, no one is: the dissidents will disappear in the crowd. Smári has scrutinized current National Security Agency (NSA) programs revealed by Edward Snowden and the overall security budget of the U.S. government, and calculated it currently costs 13 cents a day to spy on every internet user in the world. He hopes that default encryption services like his will push that closer to $10,000. It's not to stop people being spied on—he agrees that's sometimes necessary—but rather to drastically limit it. At this inflated cost he estimates the U.S. government would only be able to afford to keep tabs on around 30,000 people. "If we can't trust the government to do only those things that are necessary and proportionate—and we can't—then economics can force them to." But the reason everyone doesn't use encryption is because it's complicated and time-consuming to set up, he explains. Gmail, by contrast, is supremely sleek, simple, and fast. So Smári and two colleagues decided to develop their own easy-to-use, encrypted email system—and raised $160,000 in August 2013 from supporters on Indiegogo to do so. It's called Mailpile. "It will be feature-complete, and easy to use," Smári explains, opening his laptop to give me a sneak preview. It certainly looks good.

In 2013, documents released by Edward Snowden alleged that the NSA, working with Britain's GCHQ, and others, was—among other things—tapping seabed "backbone" internet cables, installing backdoor access to private company servers and working to crack (and weaken) encryption standards, often without much legal basis, let alone a public debate. Fearful of government surveillance, ordinary people are taking measures to make themselves more secure online, and using software designed by people like Smári to help them. The result is what I call "The Snowden Effect." More and more people are starting to adopt encryption technology—the demand for services like Mailpile, PGP, or Jitsi is growing: the daily adoption rate of PGP keys tripled in the months following Snowden's revelations. Anonymous browsers like "Tor" are becoming ever-more popular: there are now an estimated 2.5 million daily users. Facebook users, who used to be happy sharing everything with anyone, are inching towards more private settings. In the mid-1990s, the cypherpunks frequently warned of the impending "surveillance state." It turned out they were right all along. And today cypherpunk is going mainstream—thanks to a tweet.

AIN'T NO PARTY LIKE A CRYPTO-PARTY!

In 2012, the Australian parliament passed a Cybercrime Legislation Amendment Bill, which gave the government more power to monitor online communications, in the face of opposition from civil liberties groups. In the immediate aftermath, one user posted a tweet on the timeline of the Australian privacy activist Asher Wolf: "ain't no party like a crypto apps install party." A few minutes later Wolf replied: "I want a HUGE Melbourne crypto-party! BYO devices, beer & music. Let's set a time and place :) Who's in?" She later recalled that "by the time I'd had a cup of tea after tweeting the idea—I came back to the laptop and found Berlin, Canberra, and Cascadia had already set dates. By the next morning, half a dozen more countries were calling for crypto-parties."

It is second nature to people like Amir, but most people don't know how to browse the net anonymously using Tor, how to pay with Bitcoin, or how to send a message encrypted with PGP. A crypto-party is a small workshop to show them how. It's typically twenty or so people being walked through the basics of online security by volunteer experts, free to attend, and often held in someone's home, at a university, or even a pub. Wolf's tweet sparked a global, grassroots movement.* There is even a free crypto-party handbook, which was crowdsourced in less than twenty-four hours by activists all over the world, and continues to be publicly edited and updated.

Shortly after the Snowden revelations, a group of privacy activists held a very large crypto-party on the campus of Goldsmiths,

* I've documented at least 350 publicly announced crypto-parties around the world since 2012, on every continent, with anything from 5 to 500 people taking part.

University of London. I joined around two hundred people, all of whom wanted to learn how to stay anonymous online. In packed workshop sessions, each one hour long, we learned how to use Tor to browse anonymously; how to spend Bitcoins; how to use PGP. There was an interesting mix of participants. A group of older women were delighted at sending messages to each other using PGP (which is weirdly satisfying). Soon we were exchanging missives. With only a click, this:

Jklr90ifjkdfndsxmcnvjcxkjvoisdfuewlkffdsshSklr9jkfmdsgk,nm3in
j219fnnokmf9n0ifjkdfndsxmcnvjcxkjvoisdfuewlkfJflgmfklr90ifjkdf
ndsxmcnvjcxkjvoisdfuewlkf,nm3inj219fnnokmf972nfksjhf83kdbg
fhydid89qhdkfksdfhs8g93kkkafndhfgusdug892kmgsndu19jgwdnn
gskgds8t48senglsdpss9sy31bajsakf7qianfkalhs19jaslfauwq8qoafall
2kjhagfasjf993hamfalsfuqiejfkallnjksd732j1ls0dskj

suddenly becomes:

Hello!

I met a journalist worried about his sources in a dangerous part of the world overseas, and a few students who seemed happy to have found a cause to rage against. One German woman told me that she was old enough to remember the Stasi, and is convinced that we are sleepwalking into some kind of Orwellian dystopia. "Do you trust the police?" She glared at me. Well, yes, most of the time, I replied. "Well, you shouldn't!" she snapped. I asked her if she'd ever heard of Tim May and the cypherpunks. She hadn't. In fact, no one had. But so what? Surveys consistently show that we value privacy; nine out of ten Britons say they would like more control over what happens to their personal data online. The balance societies endeavor to find between individual freedoms and state power is always in flux. Most of us accept that, even in democracies, we need to be spied on sometimes—but that it should be limited, proportionate, and not misused. We pass laws to try to ensure that's the case: but modern technology has moved on so quickly, and with the advent of extremely powerful computing and the fact we share so much publicly about ourselves, a lot of people—not just cypherpunks—think that their right to privacy is being breached.

THE DOWNSIDE

People like Phil Zimmermann or Smári are developing crypto because they believe their work helps guard civil liberties from intrusive surveillance, especially in repressive regimes. And undoubtedly it does. But it's not only freedom fighters and democratic revolutionaries that use their tools. Terrorists, extremists, serious organized criminals, and child pornographers, denied mainstream channels, are often early adopters of new technology and also have an incentive to stay secret and hidden. The major producers and distributors—although not viewers—of child pornography are expert users of crypto. Without Bitcoin, the online drug market Silk Road would probably never have existed.

David Omand, the former GCHQ Director who is now a visiting professor at King's College London, remains close to the intelligence agencies in the UK. "It is absolutely vital that intelligence agencies retain the capability to monitor who they need, in order to keep the public safe," he tells me. "The internet gives a much wider range of options for avoiding surveillance. It is generally true that terrorists and serious criminals will and do use the latest technology available to them, and follow very closely the latest development in secure communications. It's an arms race." It has been alleged—although never proven—that the 9/11 terrorists used PGP encryption in their communications: "I have no idea whatsoever about that," says Omand. But he is convinced that terrorists would have been "delighted" by information about the Edward Snowden leaks. "You can be sure that they were following the story very closely indeed: as would have been the Russian and Chinese governments." The terrorist group Islamic State are known to keep close tabs on the latest developments in cryptography, using any software possible to evade monitoring.

I asked Omand if he was worried about the rise in cryptoparties, or more widespread adoption of Tor, Mailpile, and Dark Wallets. Might it make us less safe? "Yes, it does concern me. But you won't stop the intelligence machine." He thinks intelligence officers will find a way around it—they have to—but it might end up being more intrusive than using the alleged methods exposed by Edward Snowden. He recounts that during the Cold War, Soviet cyphers were too strong for GCHQ to break, so British intelligence switched to recruiting more Soviet agents. If the state considers you to be a legitimate target for security investigation but can't track your online activity using an anonymous browser, they'll put a bug in your bedroom instead. He predicts more agents and intrusive operations in the future, "which is typically more morally hazardous."

For the cypherpunks, the fact that criminals use encryption is an unfortunate outcome, but a cost worth paying for the extra freedom it provides. Zimmermann has been asked repeatedly how he feels that the 9/11 hijackers might have used software he designed. It was, he says, far outweighed by the fact PGP is "a tool for human rights around the world . . . strong crypto does more good for a democratic society than harm." Zimmermann or Tim May don't have responsibility for keeping the public safe, and don't read top-secret security briefings. Omand did. Not that he blames Zimmermann—"it is not a moral consideration for him to weigh up. Of course he should have developed PGP. We would not have the benefits of the internet without such breakthroughs. But it's for elected, democratic governments to decide whether new technologies also pose dangers for the public and what if anything needs to be done to keep down those risks to acceptable levels."

THE GULCH

In the early days, crypto was a libertarian dream—a way to spark a revolution. The cypherpunks were hard-nosed Ayn Rand libertarians, mainly concerned about individual liberty. Today, the issue of privacy and anonymity online has become a major preoccupation for people across the political spectrum. "Politically, the cypherpunks are all over the place now," says May, a little mournfully.

The majority of cypherpunks working on ways to evade state detection are not free-market warriors or convinced Randians like Tim May. Smári is a thoughtful anarchist, someone who supports the abolition of the state like May, but believes that humans, when left alone by powerful interests, will tend to cooperate and create flourishing societies, not isolated retreats. And unlike May, people like Smári worry about welfare, minority rights, and other progressive causes. But they all share a distrust of governments and centers of power—especially the security establishment—and see crypto as a mathematically guaranteed way of rebalancing democracy towards ordinary folk. Enric Duran, an avowed anti-capitalist, sees Bitcoin much like Tim May does—"an important gambit in the path which heads towards our final objective of integrated cooperatives," he tells me, via email. A world free of nation states. Crypto-currencies can "help stop our dependence on the Euro—and reduce the state's ability to control us."

Although representing radically different world views, they all believe that anonymity and privacy are vital to healthy, functioning, free society. For the cypherpunks, whether anarchist or libertarian, anonymity is about preserving the ability of people to hold multiple personalities and identities. By providing that, crypto extends the degrees of freedom individuals have, which in the long run will encourage people to live more productive and self-reliant lives, and leave more space for new ways of living. That's how Amir sees it. "This is about trying to carve out a space of freedom," he explained, "so people can do things that are worthwhile. Far better to build trust networks that are based on establishing a relationship rather than judges, bureaucracy, the police." Amir is full of ideas. Next year he plans to build industrial machines that can be used to create sustainable agriculture and waste management systems: "We're going to have our own industrial economy," he says. He thinks he'll be able to build a house for 1,000 euros and sell it for five times that, which he'd reinvest in making another Calafou somewhere else: "If they want us to play their stupid economic game, we'll beat them at it—and buy the world back." (There is already an occupied forest about an hour from London. Amir describes it as an "autonomous federation" of around forty houses.)

But if everyone starts using Bitcoin, government's ability to tax and spend will diminish: healthcare, education, and social security will suffer. The things that hold democracies together, and provide support for the most in need. Societies cannot be broken and fixed like computer code, nor do they follow predictable mathematical rules. If genuinely anonymous communication becomes the norm, it's inevitable that it will be used by criminals too. Some of the progressive groups and individuals that are fighting for digital anonymity are doing it for good reason. They don't realize they are also pushing the political agenda of a hard-line, radical libertarian from California.

Tim May doesn't care what's propelling it, because he thinks the endgame is inevitable. He tells me the third leg of the trifecta is in place: along with PGP, and anonymous browsing, there is now an anonymous currency: "And, man," he exclaims excitedly, "that's got to be freaking Big Brother out!" May anticipates that in the coming decades, governments as we know them will disintegrate—to be replaced by a digital "Gulch," something he calls a "cyberstead." A place where citizens can exist with no state at all, creating online communities of interest and interacting directly with each other. Like Amir, he's well aware that in the short term it will be turbulent for those at the bottom, even if the long-term prospects are good. "Crypto-anarchy means prosperity for those who can grab it, those competent enough to have something of value to offer for sale," he wrote in 1994. He hasn't lost his radical edge: "We're about to see the burn-off of useless eaters," he tells me, only half joking. "Approximately four to five billion people on our planet are essentially doomed: crypto is about making the world safe for the 1 percent." The short term has to be tough, he believes. It's only by removing the crutches we depend on to protect us—rules, laws, welfare—that we can grow to fulfil our potential.

I left Calafou admiring what Amir and others are attempting but worried about what it might lead to. Amir is different from May in many ways. He believes places like Calafou offer a better alternative to other ways of living for everyone—not just for the top 5, or even 1, percent—which is why they will win out in the end. But like May, he believes crypto will make this happen, without thinking precisely how, or what the consequences could be. A mathematical formula that will, with the relentless certainty of numbers, create a world of Calafous: small collectives that are self-sustaining, self-governing, owned, and controlled by the people.

At Calafou there are "people's assemblies," where the residents meet to agree on common tasks, projects, responsibilities, and so on. It's a sort of Greek agora: a method of collective decision-making that tries to involve everyone in this nascent little community. "Us in the hacker space don't bother with it," says Amir. "We don't believe in it. We want to promote individual freedom. If you have an idea, just get on and do it." As I was leaving Calafou, crossing back over the concrete bridge to the outside world, Amir told me: "There are so many people just complaining and doing nothing about it. We actually make things. We solve problems." Cypherpunks write code. And he wandered back into his very own Gulch.*

* In 2014, Taaki left Calafou. The last I heard, he was living in an anarchist squat in London.

4

THREE CLICKS

Tor Hidden Services are not easy to navigate. In many respects, they are very similar to websites on the surface net. But they are rarely linked to other sites, and the URL addresses are a meaningless series of numbers and letters: h67ugho8yhgff941.onion rather than the more familiar .com or .co.uk. To make matters worse, Tor Hidden Services frequently change addresses. To help visitors, there are several "index" pages that list current addresses. In 2013, the most well-known of these index pages was called the Hidden Wiki. The Hidden Wiki looks identical to Wikipedia, and lists dozens of the most popular sites in this strange parallel internet: the WikiLeaks cache, censorship-free blogs, hacker chat forums, the *New Yorker* magazine's whistleblower drop box.

In late 2013, I was browsing the Hidden Wiki, searching for the infamous dark net market Silk Road. As I scrolled down, I suddenly spotted a link for a child pornography website. I stopped. There was nothing strikingly different about it—a simple link to an address comprised of a string of numbers and letters, like every other website listed here. For a while I sat frozen, unsure what to do. Close down my computer? Take a screenshot? I contacted the police.

The internet has radically changed the way child pornography is produced, shared, and viewed. According to the United Nations, child pornography (which some specialists prefer to call child abuse images) refers to "any representation, by whatever means, of a child engaged in real or simulated explicit sexual activities or any representation of the sexual parts of a child for primarily sexual purposes." Most countries have different degrees of obscenity, typically based on something called the "COPINE" scale.

Once I'd opened my Tor browser, it took me two mouse clicks to arrive at the page advertising the link. If I had clicked again, I would have committed an extremely serious crime. I can't think of another instance where doing something so bad is so easy.

We can now share files and information more simply, quickly, and inexpensively than ever before. On the whole, that is a very positive thing. But not always. Is child pornography really this accessible? What does it mean if it is? Who is creating and viewing it? And in an age of anonymity, is it possible to stop them?

A HISTORY

The prohibition of child pornography is a surprisingly recent phenomenon. During the sexual liberation movement of the late 1960s and '70s child pornography was openly sold over the counter in some countries—most notably in Scandinavia—and in certain U.S. states; an interregnum that is now referred to as the "ten-year madness." By the late seventies many governments started passing tougher legislation to stamp it out, and by the late 1980s, child pornography had become very hard to find. The biggest-selling child pornography magazine in North America had a circulation of approximately eight hundred, and was distributed via a handful of shops to small, close-knit networks of dedicated collectors. In the UK, many pedophiles would travel overseas in order to smuggle it in. Law enforcement agencies in the United States considered the matter more or less under control. In 1982, the U.S. General Accounting Office reported that: "As a result of the decline in commercial child pornography, the principal Federal agencies responsible for enforcing laws covering the distribution of child pornography—the U.S. Customs Service and the U.S. Postal Service—do not consider child pornography a high priority." In 1990, the NSPCC estimated there were 7,000 known images of child pornography in circulation. Because it was so hard to come by, the numbers that were accessing it were vanishingly small. It required effort and determination, which limited it to the most motivated individuals. Even during the "ten-year madness," you didn't—you couldn't—just stumble across it.

The arrival of the internet changed everything. By the early nineties, the opportunities of networked computing were quickly exploited by child pornographers as a way to find and share illegal material. In 1993, Operation Long Arm targeted two Bulletin Board Systems that were offering paid access to hundreds of illegal images. Anonymous Usenet groups alt.binaries.pictures.erotica.pre-teen and alt.binaries.pictures.erotica.schoolgirls were both used to share child pornography in the late nineties. In 1996, members of a child abuse ring called the Orchid Club were committing and sharing live abuse using digital cameras connected directly to computers in the United States, Finland, Canada, Australia, and the UK. Two years later, the police uncovered the Wonderland Club, which comprised hundreds of people in over thirty countries who were using powerful encryption software to secretly trade images over the net. Prospective members had to be put forward by existing members, and possess at least 10,000 unique child pornographic images to join. In total, the police uncovered 750,000 images and 1,800 videos. Seven UK men were convicted for their role in the network in 2001.

As more countries went online, new production hubs sprang up. The infamous Lolita City in the Ukraine flooded the net with half a million images in the early 2000s, before it was shut down in 2004—although two leaders of the agency were taken into custody and then released.

By October 2007, Interpol's Child Abuse Image Database—made up of images seized by the police—contained half a million unique images. By 2010, the UK police database, held by the specialist Child Exploitation and Online Protection Centre (CEOP), stored more than 850,000 images (although they have since reported finding up to two million images in a *single* offender's collection). In 2011, law enforcement authorities in the United States turned over twenty-two million images and videos of child pornography to the National Center for Missing and Exploited Children.

Twenty-five years on from the NSPCC's estimate, there are today huge volumes of child pornography online, easily accessible and efficiently distributed. Between 2006 and 2009, the U.S. Justice Department recorded twenty million unique computer IP addresses who were sharing child pornography files using "peer-to-peer" file-sharing software. CEOP believes that there are approximately 50,000 people in the UK today sharing or viewing indecent images of children.

It turns out that I wasn't alone on the Hidden Wiki. According to hackers that took control of the Hidden Wiki over one three-day period in March 2014, 100,000 other people (although it's impossible to know where in the world they were based) had also visited the index, and one in ten of them had visited the link that I'd seen. According to the same source, between July 29 and August 27, 2013, there were thirteen million page impressions on Tor Hidden Services—and 600,000 of them were visits to the child pornography pages: the most popular group of pages after the index page itself. According to a 2014 study by academics at the University of Portsmouth, in the UK, child pornography sites accounted for nearly 83 percent of traffic on Tor Hidden Services.* Given the scale it is not surprising that there is no such thing as a "typical" consumer of child pornography. Although there are some broad trends—nearly all are men and often well educated—they come from all walks of life. One academic has recorded nine different types of offender, including the "trawlers" who seek out images, the "secure collectors" who obsess over secrecy and build large collections, and the "producers" who create images themselves and disseminate them. Many of these offenders would have sought and collected illegal images before the advent of the internet: it's just that the net is now the most convenient place to do

* (That data is disputed: these illegal pornography sites themselves only account for 2 percent of the 45,000–60,000 sites available through Tor, and "traffic" also refers not only to individuals but to automated "bots," DDOS attacks, and law enforcement officers who monitor the sites.)

it. But there is now another type of offender, one unique to the internet age: the browser.

THE BROWSER

"I have absolutely no idea how this happened, I really don't. In fact, I don't even understand me entirely." Michael* seems genuinely bewildered as he explains to me how he was recently convicted of possessing almost 3,000 indecent images of children on his computer. Although most of the material was categorized as "Level 1"—the least serious category, which is erotic posing but no sexual activity—his collection stretched into the more serious and obscene Levels 2, 3, and 4, and most was of girls aged between six and sixteen.

Michael is in his fifties, smartly dressed and clean-shaven. He strides confidently into the room and greets me with a friendly handshake. Until recently, he had a busy job at a medium-size business just outside the UK's second largest city, Birmingham. A married man with one grown-up daughter, a football fan who enjoys an active social life. "A very ordinary, heterosexual bloke," he tells me. "I was never—*never!*—remotely curious about young girls. It never even crossed my mind." He started watching pornography occasionally in his twenties, and dipped in periodically in his thirties. "But it was only in my forties that I started watching pornography online habitually, for sexual relief." He claims that the death of a close friend and a flagging sex life provided the prompt, but that his habit was nothing out of the ordinary. Apart from a preference for teenage girls. "I was just attracted to

---

* Not his real name.

the youth; younger faces, younger bodies. I just find teenage girls more physically attractive than women my age."

The slightly unsettling truth about sexual desire is that the law and social preferences don't neatly converge. In the UK, although the legal age of consent is sixteen, any pornography that includes someone under the age of eighteen has been illegal since the Sexual Offences Act of 2003. Yet there is a very significant and sustained demand for pornography featuring female teenagers. "Legal teen" content has always been the most competitive and populated niche market in the adult industry. According to the Internet Adult Films Database, a central repository of commercial adult films online, the most common word in film titles is "teen." In 2013, two American academics, Ogi Ogas and Sai Gaddam, analyzed almost fifty million sexual search terms that internet users had made on a popular search engine between 2009 and 2011. One in every six related to age, and the most popular by far was "teen/teens," followed by "young." Ogas and Gaddam also collected instances where a specific age was included in the search. The three most commonly requested ages that men search for online are, in order of popularity, thirteen, sixteen, and fourteen.

Sitting alongside this vast "legal teen" content is an enormous gray area of pornography known to experts as "pseudo child pornography" (although the sites themselves usually call it "jailbait" or "barely legal" pornography), which features teenagers that are, or appear to be, around the ages that Ogas and Gaddam found. The reason it's a gray area is not because the law is unclear, but because it's extremely difficult to determine how old teenagers are, especially as some try to look younger and others try to look older. The Internet Watch Foundation (IWF) is a UK-based organization that works with the police and internet service providers to try to remove online child pornography. It was set up in 1996, after the Metropolitan Police told UK internet service providers to close down around one hundred Usenet groups that they suspected of sharing child pornography. The providers proposed the IWF as a system of industry self-regulation. Every day the IWF receives dozens of reports from people who have come across what they suspect might be illegal content online. On receiving a report, each analyst carefully studies the URL content to determine whether the site contains images or videos that are likely to be illegal. The analyst will attempt to grade the severity of the material into one of the five levels. If it is judged to be illegal, the analyst will alert the police and contact the internet service provider or site administrator and request that the material be swiftly removed. If it's based in the UK, the IWF can usually get material taken offline within an hour. If the site is hosted overseas, which it nearly always is, they will do their best to work with the local internet provider or police to get it removed. They also maintain a list of blacklisted URLs to help internet service providers keep the material off. As a rule of thumb, however, the IWF can only process a referral if they believe the subject of a photograph or video is aged fourteen or under. No one knows precisely how much of this jailbait material there is, but according to Fred Langford, Director of Global Operations at IWF, they have received an increasing number of reports of it over the past ten years.

Langford claims that it is surprisingly easy to get from legal to illegal pornography simply by following website links and pop-ups. Click a link on a legal site—such as the sprawling set of "Tube" video sites—to a slightly more shadowy teen page; this in turn offers a link to a "jailbait" page; and there you might be offered yet another link . . . In this way, jailbait pornography acts as something of a gateway, both metaphorically and practically. According to research conducted by the charity the Lucy Faithfull Foundation, nine out of ten internet sex offenders did not intentionally seek out child images, but found them via pop-ups or progressive links while browsing adult pornography.

It is extremely difficult to verify these accounts. It might be an attempt by an offender to distance himself from his crime. But this is, Michael claims, exactly what happened to him. He began to visit teen pornography sites more and more regularly. And whenever he clicked on a new link—especially a free site—he provoked a "pornado" of other unrequested sites opening on his computer, caused by "pop-up" or "pop-under" sites and advertisements. These pop-up sites offered him an almost infinite array of fetishes and fantasies—and he was drawn to the jailbait categories, girls of perhaps fifteen or sixteen. He started to click.

Michael says that, after a while, he found he was spending more time in the jailbait category, and less in adult, mainstream porn sites. He never used Tor, or encryption software—his searches were all on the surface web. But he started to save and keep the images or links to sites he'd found. He felt guilty after masturbating—but never enough to delete what he'd found. They were under eighteen, but they weren't children, he says. He struggles to explain what happened next. "I can't tell you precisely when it happened, although I absolutely accept there was a point when I did cross a line." He had moved from viewing photographs and videos of teenagers, to images that, he says, were clearly of children. "It happened in tiny increments," he goes on. "I really don't remember when I moved from teens to children. But I did."

Several academic studies have examined the links between "teen" and child pornography. According to Professor Richard Wortley, Director of the Jill Dando Institute of Security and Crime Science at University College London, a lot of men get as far as the jailbait category. Some view websites once and never return. Others dip in at irregular intervals as the mood takes them. But for some, like Michael, it stimulates a sexual desire towards children of younger and younger ages. Another academic study found that adults exposed to jailbait pornography make stronger links between youth and sexuality. When Ogas and Gaddam reviewed the data they'd collected on sexual search terms, they discovered to their surprise that many people looked for taboo subjects, like incest and bestiality. The authors suggest that this is because forbidden acts also have the power to arouse, but that it's more of a psychological than a physical stimulus. Michael tells me that each time he reached one taboo, he would search for another one. "Sometimes I would see an image or website and shut it down immediately thinking how awful it was, but it remained in the back of my mind," he says.

On one occasion, around three years after his first encounter with jailbait pornography, he clicked on a pop-up that led him to a site featuring two files that he downloaded and saved. One was a video of a male adult having penetrative sex with an eight-year-old. "I remember thinking at the time that it was terrible. That I never wanted to look at it again. But I kept it, just in case."

Michael considers himself to be a deeply moral man, and repeats several times that he'd never harm or hurt anyone, especially not a child. "It didn't seem real," he tells me. "I accept this is a false distinction now, but in the videos or images, they always looked like they were not being harmed. I made excuses in my head as to why it was okay. For a while I told myself what I was doing wasn't even illegal."

BOY-LOVERS AND COGNITIVE DISTORTION

Elena Martellozzo is an Italian academic who has worked closely with CEOP. She explains to me that offenders like Michael often claim there is a distinction between the real and the digital, and will even construct wild justifications to convince themselves that their online behavior is somehow acceptable. One important aspect of John Suler's famous Online Disinhibition Effect is the "dissociative effect"—the idea that a screen allows you to disassociate your real self from your online behavior, to create fictitious identities and alternative realities, in which social restrictions, responsibilities and norms do not apply: as if the online space is somehow separate and different.

This dissociative imagination is most visible in several online pedophile communities. Users of these "legal-only" forums congregate to speak openly about their desires, without posting or sharing any illegal material. There are a number of these hosted around the world—both open and closed—often with several thousand members and visitors. Far from being uncomfortable or secretive about their desires, in legal-only forums pedophilia is celebrated, with users proclaiming it to be a misunderstood, yet perfectly natural condition. One site I found—which appeared to be United States–based although it's hard to be sure—offers "mutual support among child-lovers who are sexually attracted to boys." Members of this group openly and proudly discuss their attraction, using an array of board-specific terminology: AF (adult friend), AOA (age of attraction), BM (boy moment—an experience that an adult has had with a boy in his everyday life). "Posters," reads the forum rules, "have the ability to relate to boys in a magical way."

According to Elena Martellozzo, these sorts of rituals allow boy- and girl-lovers to construct a board-specific alternative reality. Some of these forums establish complex and fantastical hierarchies, run according to exacting rules for building trust and progressing within the group. One forum studied by Martellozzo called the Hidden Kingdom* had a pyramidal structure based on medieval titles: the Lord was the overall site domain owner, while Knights of the Realm and the Royal Inner Circle had the power to moderate the forum and ban people who'd posted illegal material. To climb the hierarchy, Townsfolk had to post on the site at least fifty times a day. In many ways, these legal-only forums are just like any other: hierarchies, in-jokes, and memes, complaints and frustrations. In another legal-only forum, I recorded grumbling about trolling from one regular poster; a complaint about the "selfish" users who don't give feedback on uploaded videos on a child pornography site; and several discussions about how law enforcement agents and child protection agencies are persecuting them for their natural and healthy love towards children.

The most well-known of these groups, which has been in operation since long before the Web, is called the North American

* Which has since been taken offline and at the time of writing is currently being investigated by the Metropolitan Police.

Man/Boy Love Association (NAMBLA). Founded in the United States in 1978, their goal is to "end the extreme oppression of men and boys in mutually consensual relationships," although in truth it's hard to see how that really amounts to anything except the right of their members to fulfil their wish of having sex with children. NAMBLA members consider themselves to be misunderstood and persecuted in the same way homosexuals once were, and in the 1980s and '90s they held public demonstrations in support of their cause. Most remarkably NAMBLA considers itself part of "a historic struggle" and claims to support "the *empowerment* [my italics] of youth in all areas" against what they call "rampant ageism."

VIRTUAL AND REAL ABUSE

The relationship between these "virtual abuses" and child abuse in the real world is not clear. Notwithstanding the gravity of possessing child pornography, academic studies that have examined the causal link between viewing material and physically abusing children are inconclusive. For some men, watching child pornography might spark an interest that will lead them to try to contact children. For others, a sexual interest in children remains a fantasy that they would have no intention of ever enacting. Many internet sex offenders like Michael say they would never offend "in real life" and even give ethical or moral reasons for not doing so. For others, viewing material may even provide an outlet that prevents them from moving on to real-world abuse.

While there are more people being convicted for possession of child pornography now than twenty years ago, there has not been an increase in the amount of recorded physical sexual abuse. In fact, in the United States, data aggregated from state child protection agencies pointed to a drop of 62 percent since 1992, while in the UK the numbers have remained stable since the mid-1990s (although with an increase in emotional abuse). According to the American academic danah boyd, every new technology results in new anxieties about young people's safety, but it's often not borne out by the evidence. Despite fears about online predators, the vast majority of victims are abused by someone they know—fathers, stepfathers, or another family relation or family friend.

But the internet has changed the way certain types of sex offenders operate. The police have recorded an increase in the proportion of grooming cases that involve an element of online interaction, and are worried by the way groomers trawl through social networking sites actively seeking out vulnerable young people. According to Peter Davies, the former Director of CEOP, the "internet amplified, multiplied and in some cases almost industrialized [grooming] to a quite remarkable degree." At the same time, patterns of abuse conducted over social media are changing. A CEOP and University of Birmingham study showed that physical contact was a declining motivation in online sexual abuse of children: there is a drop in online grooming to meet children offline, and an increase in the amount of purely online abuse. In one sting operation led by the Metropolitan Police, a fake social networking profile was visited by 1,300 people, with 450 adult male profiles initiating contact. Eighty of them became virtual friends with prolonged communication via private chat, and twenty-three of them became involved in abusive sexual behavior.

Tink Palmer is uniquely qualified to explain how the net has changed grooming. She is the Founding Director of the Marie Collins Foundation, a charity which helps victims of sexual abuse. When Tink first started working in the field, pre-internet, the accepted model of grooming was called the "Finkelhor Model." It describes grooming for sexual abuse as a four-phase cycle. First, there is the motivation stage, when the abuser develops the desire to act. The second phase requires overcoming internal inhibitions—the emotional and moral qualms he or she might have. Once justified, he or she must also overcome external inhibitions: family members, neighbors, peers, locked doors. The fourth and final stage involves overcoming the resistance of the victim.

"When I started," explains Tink, "grooming was a relatively slow and careful process. A groomer would try to gain access to a young person, often by befriending the family or becoming part of their wider social circle. They would slowly try to build up a rapport with the child, and subtly try to turn the relationship towards sex, before drawing them into compromising situations." Tink does not think the internet has changed the model—it's still a cycle of abuse—but the dissociative effect of communicating behind a screen has dramatically sped the whole process up, and reduced the external inhibitions—the physical barriers that make access to children difficult. Groomers still have to build up a rapport, explains Tink, just as they do offline. But they do it with new technology. "They spend hours monitoring their victim's social networking profiles to learn about them, and then use that information—favorite movies, places they've been, a recent status update—to try to create a rapport." They will, she says, learn text and internet language and behavior, and all the abbreviations that come with it, and she reels off dozens of expressions that I more commonly associate with teenagers: "do you have pos atm?" (parent over shoulder at the moment?); "tdtm" (talk dirty to me), and so on. Has it made grooming easier? I ask. "Definitely!" says Tink. "It's quick. It's anonymous, it disinhibits. Groomers haven't got to leave the house to sexually harm children."

Many online groomers are extremely cautious, and give few or no details about themselves until they feel certain they are chatting to a real child. But the dissociative effect has made some groomers feel more disinhibited in what they say, which allows them to more quickly open up what Tink calls "the sexual phase." It's that critical moment when a groomer brings up the subject of sex with the potential victim, and usually when they have already isolated them in private chats via MSN or direct messaging. In her careful study of twenty-three Metropolitan Police investigations into online grooming, Martellozzo found many are surprisingly public and open in their intentions. One had even posted naked photos of himself, while another was using a public social media profile which read, "I am a nice, decent, very loving caring guy with a pervy side—daddy/daughter, incest etc." Tink has also seen the groomers' targets change their patterns of behavior. If the traditional online grooming involved a very slow building of rapport, a growing number of cases now involve the *victim* opening up the sexual phase. It's not to excuse the behavior of the adult in any way—but Tink thinks it reflects young people becoming socialized to what behavior they think is expected of them online. She tells me of one recent case involving a fourteen-year-old girl who, on first contact with a groomer in his twenties, told him that she was ready to have sex with him. Tink has seen young girls dangling older men along for fun or to relieve boredom online (it is, she says, called "bag a pedo"). But it's not always quite as safe as young people might think.

STEMMING THE TIDE

A typical day for an analyst at the IWF is never easy. Twelve of them—a mix of men and women that includes a former fireman and a recent graduate—work at a nondescript business park just outside Cambridge. When I arrive, early one cold February morning, only a small printed A4 page with the words "Internet Watch Foundation" on the door offers a hint as to what happens inside. The offices are modern: spacious, bright, open-plan. A radio hums in the background and the staff cheerfully buzz about. Four walk into a secure room that's about the size of squash court. Inside, a large peace lily hugs the far wall next to a Banksy painting. The only thing that sets these rooms apart from thousands of other twenty-first-century offices is the absence of family photos on the desks.

On his very first day, a decade ago, Fred Langford went through the painful ordeal that all new staff at the IWF are subjected to: viewing the images and videos in ascending order of obscenity, from Level 1 to Level 5. It's the final test for the job, really the only way of knowing whether new recruits will have the measure of it. At the end of the session, which is usually held on a Friday, they are told to take the weekend to make up their minds about joining. After seeing Level 1, Fred recalls thinking, "Oh, this isn't so bad!" By the time he'd gone through the whole spectrum, he'd changed his mind. "As I cycled home, I had this Level 5 image going round and round in my head. I couldn't get rid of it." He told his partner he couldn't take the job after all. But by Sunday evening, like most new recruits, he'd changed his mind. "I decided I wanted to help—to do anything I could to stop this." The IWF is a—perhaps *the*—world leader in trying to stem the tide of illegal pornography on the net. It's a grim task.

In 2013, analysts at the IWF documented around seven thousand URLs that contained scenes of torture and rape, often of children under ten. It's hard to imagine how anyone can maintain their sanity in these circumstances. That's why all staff, even the head of media, are put through a rigorous annual psychological examination. They are encouraged to take breaks whenever they need, leave early, and have monthly counseling. But it's still difficult, even for experienced analysts. Everyone here does what they can to keep their private and professional lives separate. That's why there are no family photos on the desks. "Each one of us has our own personal coping mechanism," says Fred.

Staff at the IWF must sometimes feel like King Canute, trying to hold back a tide of reproducible, shareable files from appearing online. When the IWF was founded in 1996, its job was to target 100 illegal newsgroups. By 2006, that number had risen to just over 10,000. In 2013, it was 13,000. The IWF and the police are frequently faced with new challenges. In 2013, the IWF started receiving dozens of complaints about the same website, but every time they checked the URL, all they saw was mainstream adult material. After careful forensic work, the IWF's technical researcher discovered that if you'd arrived there via certain other sites in a certain order, a trigger would kick in and send you to a hidden version of the same webpage. This is known as a "disguised cookie site."

Despite the difficulty, they have achieved some considerable successes. In 2006, the IWF registered 3,077 domains hosting child pornography. In 2013, this was down to 1,660. They have been particularly good at shutting down UK-based sites. In 1996, 18 percent of all sites were hosted here; now that's under 1 percent. It's good, but it could be better, Langford accepts. And unlike in the 1980s, cutting down on the local supply is not enough. Today an image could be created in one country, held on a server or website in another, and viewed by someone in another. Crossing international jurisdiction can be infuriating, especially when content is hosted in countries where law enforcement or internet service providers seem less concerned about the subject. Langford tells me of a "revenge porn" website that is hosted in Germany, where he estimates at least half the videos and images are of people—mostly British girls—under the age of eighteen. He's been trying to get the German internet service provider to act for weeks, but to no avail.

The overwhelming majority of the material investigated by the IWF is on the surface web, accessible with a normal browser like Google Chrome, usually hosted in countries where domestic police are uninterested, incapable, or under-resourced. Often, a link will take users to a "cyber-locker," a hacked website where files are stored without the owner realizing. Around a quarter of all referrals received by the IWF are from commercial sites, which ask for credit card payment for access and are advertised through spam mail.

The IWF doesn't investigate URLs of Tor Hidden Services. Firstly, because they receive very few reports of material on there, and secondly, because if they did, there isn't very much they could do about it. Tor Hidden Services can be hosted on any computer, anywhere in the world, and the complicated traffic encryption system used by Tor means it's very difficult to work out exactly where it is or who they could contact to take it down. But Tor Hidden Services are vital to understand why the IWF's job is so hard, because they serve as a hub that produces new material and recycles old, making it available to a wide audience to view, save, and share again (often using peer-to-peer file-sharing technology). Content on these sites is uploaded anonymously using Tor and encrypted file sharing and downloaded by other users, decentralizing and widening the distribution base of material. Every time a Hidden Service is taken down—it's hard but not impossible—the community dusts itself off, reorganizes, and starts again.

In 2011, there were believed to be at least forty Tor Hidden Services that hosted child pornography, the largest of which contained more than 100 gigabytes of images and video. The same year the hacktivist collective Anonymous—which, although generally in favor of unfettered free expression online, does object to child pornography—managed to locate the server where some of these sites (including Lolita City) were hosted, and knocked them offline in what they called Operation DarkNet. But within a few days, most were running again using different servers, and with more visitors than before. Between June 2012 and June 2013, one exploitation ring was abusing children and sharing images via a Tor Hidden Service to 27,000 visitors, hosting 2,000 videos and involving 250 victims. By June 2013, Lolita City had grown to 15,000 members, and its database had increased to well over one million illegal pictures and thousands of videos.

Next it was the FBI who struck. In August 2013, following a

lengthy investigation, they arrested Eric Eoin Marques*, the twenty-seven-year-old Irishman who, they alleged, ran Freedom Hosting, which provided server space for many of the most notorious Tor Hidden Services, including criminal hacking site HackBB, money laundering sites, and over one hundred child pornography sites. After Freedom Hosting was taken offline, most of the major child pornography sites went down with it.

Child pornography forums on Tor Hidden Services were ablaze with rumor and discussion about the arrest. They were trying to work out collectively where the new sites were, and how to access them: discussing downloading times, the quality of material, and above all the security features of new sites. Shortly after Marques's arrest, backup or "mirror" versions of the sites were up and running again on new servers. Slowly, functioning links started to reappear on the Hidden Wiki, as users began creating new servers and uploading their own collections again. Then it was the turn of a couple of lone vigilantes to take it down. In March 2014, a hacker called Intangir, together with another who uses the Twitter handle Queefy, managed to take control of the Hidden Wiki I had accessed, and close it down, along with its child pornography links. But, as this book went to print, it was all available again—as if nothing had happened.

ARREST

Although their job is difficult, the authorities still make hundreds of arrests a year. According to the Department of Homeland

* At the time of writing, Marques denies all charges—and the U.S. government is pressing for his extradition from Ireland.

Security, the agency has made 8,000 illegal pornography arrests over the last decade. Increasingly, police units around the world join forces—since they recognize that this is an international problem. In 2013, a major international investigation, led by Canadian and U.S. police, resulted in the arrest of 350 people around the world. That same year, the UK police also came knocking for Michael. He describes the moment the police presented him with a warrant for his arrest—he was at home with his wife and daughter.

That experience can prove too much for many. In 1999, the FBI seized the database of a company called Landslide Inc., which it suspected of selling child pornography on the internet. The database was found to include the credit card details and IP addresses of over 7,000 Britons, data which was swiftly handed over to the British police, who subsequently made almost 4,000 arrests. It resulted in 140 children being rescued, and in the suicides of 39 of those arrested. Although no data is available, Tink believes that suicide rates are higher among those arrested for offences online than those perpetrated in the real world. The online offenders continue to retain Suler's dissociative fantasy. "It was only when the police arrived," says Michael, "that I realized the severity of what I'd been doing."

For obvious reasons, Michael wants the machine to bear some of his guilt. "I cannot believe there is so much of it out there!" he tells me, when I ask him what should be done to stop people accessing child pornography. "Why on earth was it so easy for me to find it?" According to Professor Wortley, the potential to become sexually attracted to children is not as rare a phenomenon as we'd like to imagine. The human sexual impulse is extraordinarily flexible, and at least partly shaped by social norms. Without some degree of demand for these images, they wouldn't be produced and shared in such staggering volumes. This is why the net has led to such an explosion in both content and the number of people accessing it: by making it easier to find, the latent demand can be more readily realized and, in some cases, created.

This does not excuse what Michael did. Just because something is three clicks away does not make it any less of a crime. Michael repeated to me several times that he never actively searched for the material. He clearly thinks that mitigates in his favor. But the distinction between searching and accidentally-finding-and-keeping is pretty meaningless on the internet. Michael clicked three times: and then he kept clicking. It's not the computer's fault. It's Michael's fault. But if it had been a little harder for him to find, if jailbait pornography wasn't so easily accessible, perhaps Michael's casual or vaguely formed attraction to children would never have been explored. Without the internet, I don't think Michael would be a convicted sex offender.

WHAT NOW?

The task of ridding the internet of child pornography is exceptionally difficult. Michael is just one type of sex offender, and is at the less serious end of the scale. There are many more committed sex offenders than him, and no matter what we do, they will always search, find, and share obscene images, and the police will always try to catch them. The criminals are getting smarter, but so are the authorities. The chief task for organizations like the IWF and the police is to keep bearing down on supply as far as possible to limit the content that is available and make sure people like Michael—browsers—realize that they might get caught. The flow of material probably can't be stopped, but anything that can stem the tide can and does make a difference.

But bearing down on supply is getting harder. In addition to Tor Hidden Services, popular culture is prevailing against the IWF. According to a major review into the sexualization of teens, conducted in 2010 by Dr. Linda Papadopoulos for the UK Home Office, many young people are developing unhealthy attitudes and patterns of behavior towards sex. Pornography of all types is now widely available and easily accessible to young people—and more of them are watching it at an earlier age. This is the awkward secret of child pornography: a growing proportion of it is made by the victims. They, too, are subject to the same dissociative effect. Although data is highly variable—with estimates of teens in the United States and UK who have created a sexual image or video of themselves or sent sexually explicit messages ranging from 15 to 40 percent—the number is believed to have increased dramatically in recent years. According to the NSPCC, sexting has become the "norm" among young teens. It's quite a natural thing for young people to explore their own sexuality. But the moment a digital file is posted online, it is almost impossible to control who sees it, and what they do with it. There are sex offenders who trawl the net searching for this material, which they will find, save, and share with others. According to the IWF, as much as one third of all material they see is now self-generated, and it covers all five levels of obscenity. Digital files: reproducible and shareable at almost no cost.

After I left the Hidden Wiki, I went on to the safer and more familiar surroundings of Facebook. "Hottest Teens 2013" popped up. It read: "Teens: post your sexiest pics on this page! Whoever gets the most likes from other Facebook users will be declared the winner." Twenty thousand had already signed up.

5

ON THE ROAD

The internet has transformed commerce and trade. It seamlessly connects buyers and sellers around the world, opens new markets and makes shopping simple, convenient, and quick. Approximately 50 percent of all global consumers now make online purchases—a percentage that grows every year. But alongside the multibillion-dollar world of e-commerce with its buy-it-nows, one-click buys, and next-day deliveries, exists another market that is growing just as rapidly. In this world everything—legal and illegal—is for sale.

According to a 2014 survey of almost 80,000 drug users from forty-three countries, an increasing number of users say they have sourced their drugs online: approximately 14 percent of U.S. drug users scored from the net. And the majority of them went to one place. I don't take illegal drugs, and I've certainly never bought them before, but this morning an innocuous-looking white envelope was posted through my door. It contains a very small amount of high-quality cannabis. With a few simple clicks I'd done what approximately 150,000 people have done over the last three years: I bought drugs on the Silk Road.

In 1972, long before eBay or Amazon, students from Stanford University in California and MIT in Massachusetts conducted the first ever online transaction. Using the Arpanet account at their artificial intelligence lab, the Stanford students sold their counterparts a tiny amount of marijuana. It was the start of a small but very noticeable trend. Throughout the nineties groups of dealers would periodically pop up in online drug discussion boards to sell niche narcotics to drug connoisseurs.* By the early 2000s, the first large-scale online drug market had appeared on the surface web. The Farmer's Market offered an email-only service selling mainly psychedelics. According to an FBI indictment, between January 2007 and October 2009, the Farmer's Market processed over 5,000 orders, and a million dollars' worth of sales in twenty-five countries. Then in 2010, the Farmer's Market became a Tor Hidden Service.

Today there are thought to be between 40,000 and 60,000 Tor Hidden Service sites in operation (due to its encryption system, it's very difficult to measure accurately). With a sophisticated traffic encryption system, Tor is the ideal place for unregulated, uncensored markets. Although many Hidden Services are legal, approximately 15 percent relate to illegal drugs.

On November 27, 2010, a user named altoid posted the fol-

* This ad hoc dealing continues today, often on forums related to prescription drugs.

lowing message on the surface net magic-mushroom forum, the Shroomery:

> I came across this website called Silk Road. It's a Tor hidden service that claims to allow you to buy and sell anything online anonymously. I'm thinking of buying off it, but wanted to see if anyone here had heard of it and could recommend it.

Two days later altoid turned up on bitcointalk.org, a discussion forum about the crypto-currency: "Has anyone seen Silk Road yet? It's kind of like an anonymous Amazon.com. I don't think they have heroin on there, but they are selling other stuff." altoid linked to a Wordpress blog that gave further information: "Marijuana, Shrooms, and MDMA" were already for sale, and it urged users to register as buyers, or "join as a vendor." Word started to spread, and by spring 2011, a handful of sellers had signed up, attracting a small number of buyers. By May 2011, there were over 300 product listings, nearly all of them illegal drugs. When news of this new "anonymous marketplace where you can buy anything" was reported in the online magazine *Gawker* in June 2011, the response was predictable. Thousands rushed to join.

What these new visitors found was a radical alternative to rickety, unprofessional sites like the Farmer's Market or risky ad hoc deals through forums. As altoid suggested, the site was professionally and intuitively designed. On the left-hand side of the webpage there were categories listing the different products on offer, and, when you clicked through, photographs of each. Vendors, too, were well represented. Each was listed with a short description and contact details. A link to customer service complaints was prominently displayed, as was your shopping "cart," and how much money you had in your account. Behind the slick façade was a sophisticated security system. The site was accessible only via a Tor browser, products could be bought only with Bitcoin, and visitors were advised to sign up with digital pseudonyms. Any correspondence between buyers and sellers took place using PGP encryption, and once read, messages were automatically deleted. In June 2011, a secure forum was set up in order to enable better communication between users of the site.

As well as being customer-friendly, the site was also extremely well managed. In October 2011, altoid returned to bitcointalk .com, no longer posting as a curious potential shopper, but as a key member of the quickly burgeoning site, "looking for an 'IT pro'" to help to maintain it. At that point, a team of between two and five administrators kept the site running, dealing with buyers' and sellers' complaints, resolving disputes, and scanning the site for signs of any possible infiltration by law enforcement agencies. These administrators submitted a "weekly report" to the main site administrator—a user named Dread Pirate Roberts (DPR)—via Tor Chat and an internal email system, describing work completed and any issues that needed resolving, asking for guidance, and requesting leave. Silk Road received a cut of all the sales that went through the site, and the administrators received a salary of between $1,000 and $2,000 a week for their troubles.

Despite the occasional hack, vendor arrest, and dispute over site commission rates (most notably when the website announced key changes to its rates in January 2012), the Silk Road kept growing. According to the FBI, by July 2013 the site had processed over $1.2 billion worth of sales. Almost 4,000 anonymous vendors had sold products to 150,000 anonymous customers across the world, and DPR was believed to be making $20,000 a day on commission alone.

This was by far the most sophisticated online drug market ever seen. And it was a project that was motivated by more than financial gain. When you first arrived at the original Silk Road, a message from DPR greeted you:

> I'd like to take a moment to share with you what the Silk Road is and how you can make the most of your time here. Let's start with the name. The original Silk Road was an old-world trade network that connected Asia, Africa and Europe. It played a huge role in connecting the economies and cultures of these continents and promoted peace and prosperity through trade agreements. It is my hope that this modern Silk Road can do the same thing, by providing a framework for trading partners to come together for mutual gain in a safe and secure way.

The name Dread Pirate Roberts was taken from the 1973 book *The Princess Bride* in which the Pirate was not one man, but a series of individuals who periodically passed the name and reputation to a successor. The name was chosen for a reason. Silk Road was a movement. "We are NOT beasts of burden to be taxed and controlled and regulated," wrote DPR in April 2012. "The future can be a time where the human spirit flourishes, unbridled, wild and free!"

Across Tor Hidden Services forums—but extending out into surface net forums like 4chan and Reddit—a bustling ecosystem grew up around the Silk Road, uniting an eclectic mix of "roadies": libertarians, Bitcoin fanatics, drug aficionados, and dealers, all committed for their own reasons to the idea of an unregulated online market. This sprawling community constantly monitored the market, checked security vulnerabilities and performance, and updated others on what they found. I contacted one of the moderators who ran Silk Road's popular Reddit group before it was closed down. "It's become a sort of safe haven for people who agree that no government should be able to tell them what they can put in their own bodies," he told me. "Users and sellers alike can have the freedom to be open and express themselves in ways that are impossible in real life."

Everything changed in autumn 2013. Despite the efforts of site administrators and the Silk Road communities, undercover FBI agents had been making purchases on Silk Road from November 2011, and had been closely tracking DPR and other key vendors and site admins. On October 1, 2013, they arrested twenty-nine-year-old Ross Ulbricht in a San Francisco library on suspicion of drug trafficking, soliciting murder, facilitating computer hacking, and money laundering.* They believed that they had found the Dread Pirate Roberts.

Ulbricht was a university graduate and self-confessed libertarian who, until his arrest, had been living under the name Joshua Terrey in a small shared apartment near to the library. He had told his housemates that he was a currency trader, recently returned from Australia. The FBI alleges that they confiscated 144,000

* In February 2015, Ulbricht was convicted of seven charges, including narcotics and money laundering conspiracies; and faces, at the time of writing, a minimum of thirty years' imprisonment. "Ross is a hero!" shouted one supporter from the courtroom as the decision was announced. His legal team plans to appeal the decision. Ulbricht has suggested that another individual, called Variety Jones, was his mentor—possibly even the mastermind of the whole site. Jones remains a mystery.

Bitcoins (amounting to some $150 million) from Ulbricht's computer. There swiftly followed the arrest of several suspected high-profile Silk Road administrators and dealers in the UK, Sweden, Ireland, Australia, and the Netherlands.

Shortly after Ulbricht's arrest, visitors to Silk Road were greeted with a new message: "This Hidden Site has been seized by the Federal Bureau of Investigation." The news quickly spread. "IT JUST HAPPENED OMFG OMFG OMFG OMFG," wrote one anonymous user on 4chan's /b/ board, sharing a screen grab of the FBI takedown notice within minutes of the site being removed. "Do you guys realize what this means?" replied another. "It's not just about pedos with their pizza [a code word for child pornography] or us with our drugs. We are losing every safe haven we've got." Silk Road forums—which were still up and running, operating as they did on different servers from the site itself—were in a state of panic.

Was this the end? Not quite. Seven days after Ulbricht's arrest, Libertas, who had been a Silk Road site administrator since February 2013, reemerged on the forum and posted the following:

> Ladies and Gentleman, I would like to announce our new home . . . Let L[aw] E[nforcement] waste their time and resources whilse we make a statement for the world that we will not allow jackbooted government thugs to trample our freedom!

Silk Road had returned as Silk Road 2.0: a new, better, and safer site. Libertas predicted it would be up and running within a month. For dramatic effect, the temporary Silk Road 2.0 landing page featured a doctored version of the FBI takedown notice. Libertas and other site administrators had been working around the clock to rebuild the site using some of the source code from the original and to reinstate as many of the old vendors as quickly as possible. Although plenty of roadies were unhappy that the site had vanished along with their Bitcoins, most were desperate to get back to business. Inigo, one of Libertas's fellow administrators, complained about being inundated with emails from sellers trying to get started again: "We are going as fast as we can," he apologized on the forum.

One month later, they were ready to go. True to the name, a new Dread Pirate Roberts resurfaced to run the site. (As of writing, his identity remains a mystery.) "You can never kill the idea of #silkroad," announced the new incumbent on Twitter on the morning of November 6, 2013. He then switched over to the forum: "Silk Road has risen from the ashes and is now ready and waiting for you all to return home. Welcome back to freedom . . ." Silk Road was up and running again.

But despite Libertas and Inigo's best efforts, Silk Road had lost its market dominance. It wasn't the only dark net market, it was just the largest. Others had spotted an opportunity, and from 2012 several competitor markets began to appear, including the Black Market Reloaded and the Russian Anonymous Marketplace. The disappearance of the market leader in October 2013 heralded six months of mayhem. New markets were founded, hacked, shut down by law enforcement, and reopened again. There were dozens of spoof markets set up in order to trick buyers out of their Bitcoins. A number of Silk Road users flocked to the Sheep Market soon after the FBI takedown, but after a short period of activity, it disappeared—either hacked or deliberately taken offline—along with everyone's money. The highly anticipated Utopia marketplace was set up in the first week of February 2014, but shut down by the Dutch police within a fortnight. Buyers and vendors who'd become used to the stability and reliability of Silk Road were struggling to work out which sites to trust.

An atmosphere of paranoia and suspicion took hold. The authorities appeared to be winning. But not for long. By April 2014, the markets were settling down. Three large sites gradually emerged as both trustworthy and reliable, and began to grow—selling more products than ever before. Order was resumed. Between January and April 2014, Silk Road 2.0 alone processed well over 100,000 sales. It was as if nothing had happened.

Since the arrival of these "dark net markets," there has been—understandably—uproar and consternation. The *Sydney Morning Herald* warned of "the flourishing online drug market authorities are powerless to stop" in 2011, while in 2012, the *Daily Mail* called Silk Road "the darkest corner of the Internet." Charles Schumer, the U.S. senator who demanded an investigation into the Silk Road in 2011, described the site as "the most brazen attempt to peddle drugs online that we have ever seen." But it is not surprising that online drug markets exist. What *is* surprising is that they work. Dark net markets are uniquely risky environments in which to conduct business. Buyers and sellers are anonymous, and never meet. There are no regulators to turn to if the seller or the site administrators decide to take your money. It's all illegal, at constant risk of takedowns or infiltration by law enforcement agencies. And yet, despite these conditions, dark net markets are thriving. All of these marketplaces—the Silk Road, Agora, Evolution, or any one of the dozens of others—work in pretty much the same way. How?

ON THE ROAD

You can't access dark net markets using a normal browser. Like other Tor Hidden Services, you can only access them using Tor.* Buyers therefore tend to arrive at the sites via the Hidden Wiki, or one of the many other index pages that help you navigate this opaque world.

I've just arrived at one popular index site. The first thing I notice is how many dark net markets there are, and the dizzying variety of drugs they claim to sell. There are now at least thirty-five functioning marketplaces, and deciding which one to choose is extremely difficult. Most of us are faced with this dilemma every day.

According to Nathalie Nahai, the author of *Webs of Influence*, a study on online persuasion, we make subconscious judgments about websites based on "trust cues." Typically, explains Nahai, we have confidence in a site if it is well designed—with high-definition logos and page symmetry—simply constructed and easy to use. It is an indication of the amount of effort the people behind the site have put in, and, Nahai argues, a reliable measure of how deserving they are of our trust and custom. Major e-commerce companies spend millions developing and designing websites. Many dark net markets do the same. All use recognizable logos, and all develop unique branding. Silk Road 2.0 retained Silk Road's well-known logo—an Arab trader on a camel, all in green—when it resurfaced in November 2013. The Agora Market's logo is a masked bandit, brandishing a pair of guns. The

* In 2014, Tor2Web offered users access to Tor Hidden Services using unencrypted browsers, although, because they do not mask the IP address of the user, they are rarely used.

Outlaw Market's masthead features a cowboy. All the sites also share the same basic features: profile page, account, product listings. According to Nahai, these are trust cues too—items that customers expect to see.

Like any market, sites also compete to draw customers in. In April 2013 Atlantis, a rival marketplace to the Silk Road, ran an aggressive campaign to encourage users to switch allegiances: "You need to give customers a good reason to move from their existing market. We do this in several different ways: usability, security, cheaper rates (for vendor accounts AND commission), website speed, customer support, and feedback implementation," the site administrator explained. Each market adds its own embellishments. The Pirate Market has a neat little online gambling game of rock paper scissors, and a feedback option: "tell us what you don't like about this site."

Logos and welcome emails aren't quite enough on the dark net markets. The Sheep Market's pleasing aesthetic counted for little when the site disappeared with almost $40 million of buyers' and vendors' Bitcoins. Silk Road 2.0 was hacked in February 2014, with around $2.7 million in Bitcoins lost. To get a handle on who I could really trust, I headed for the dark net market forums. If there is a scam site or vendor operating, this is where you'll learn about them. There are dozens of Reddit threads, user-generated blogs, and specialist forums on the surface net dedicated to researching each marketplace, collating user experiences, and discussing security features. Silk Road 2.0 is still a popular choice. I read a number of posts praising the way administrators responded to the February 2014 hack. Defcon, the new site administrator, immediately promised to reimburse every vendor who'd lost money—and even claimed that the site admins would receive no commission until every dispute was resolved. By April 2014, Defcon triumphantly declared that they had paid back half of the lost Bitcoins. Silk Road 2.0 also offers the widest variety of products from the largest number of vendors: 13,000 listings, compared to the second largest, Agora Market, which has 7,400. Positive endorsements, a wide range of products, excellent security. I need no more persuading.

VENDORS AND PRODUCTS

Signing up for Silk Road 2.0 is extremely simple. Username. Password. Complete the CAPTCHA (Completely Automated Public Turing test to tell Computers and Humans Apart), and you're in. "Welcome Back!" reads the landing page.

The forums were right—I am immediately overwhelmed by choice. There are around 870 vendors to choose from, selling more drugs than I'd ever thought possible. Under ecstasy alone, I find listed: 4-emc, 4-mec, 5-apb, 5-it, 6-apb, butylone, mda, mdai, mdma, methylone, mpa, pentedrone, pills. But the choice is not limited to drugs. There are sections for Alcohol, Art, Counterfeit, and Books, and listings including a hundred-dollar Starbucks e-gift card priced at around $40, a complete box set of *The Sopranos*; a hundred-dollar Marine Depot Aquarium Supplies voucher, Fake UK Birth Certificates, Fake Gift Cards, and something called PayPal Win All Disputes—The Ultimate PayPal Guide.

The nature and volume of trade on the dark net markets have always been shrouded in mystery: after all, how would you collect the data? In early 2014, an anonymous user of Silk Road 2.0, using a clever computer program, harvested the details of 120,000 sales that had passed through the site over a ninety-nine-day period between January and early April 2014 and dumped it in a file on an obscure Silk Road discussion forum. It provides the most detailed look ever into the comings and goings of the site.

Not surprisingly the most popular products are drugs. "Weed" was the top-selling item (28 percent of all goods sold), followed by cocaine (19 percent), MDMA (18 percent), digital goods (14 percent), hash (12 percent), and cannabis (8 percent). But if a vendor offers something interesting and unique—at a good price—he or she can clear enormous volumes very quickly, regardless of what it is:

Top-Selling Items on Silk Road 2.0, January–April 2014

Product	Price ($)	No. of sales (over 99 days)
Australia Genuine Roche Valium 10 x 10mg	42	240
1000 2mg Pfizer Xanax Bars	1050	193
The Original Lotus Coupon Collection [fake gift vouchers]	84	190
Testosterone Enanthate 250 (250mg/ml) 10ml U.S.	40	187
Reality Kings Premium Account [pornography]	10	142
Australia Genuine Valium Single Tablets 10mg	5.25	117

It is a truly international market. Although vendors tend to be based in the United States (33 percent), the UK (10 percent), or Australia (10 percent), most promise to ship to every country in the world.

The market is characterized by a small number of very large dealers, and a long tail of more moderately sized operations. Twenty-one vendors sold over 1,000 items between January and April 2014; while 418 sold fewer than 100. (The most active seller made 3,592 sales over the period.) The typical seller (the average of our sample of 867 vendors) sold 178 items. According to data released by the U.S. government in 2014, at its peak, Silk Road 2.0 had 150 thousand active users—and was generating sales of around $8 million a month.

By analyzing the sales data of the most active dealers cross-referenced against the value of each product, I was also able to calculate a rough estimate of the sort of turnover the top dealers make here.

Turnover: Top Vendors on Silk Road 2.0, January–April 2014

Name	Products	Overall turnover (99 days)
The Drug Shop	Principally cocaine, heroin, and ketamine	$6,964,776
Heavenlost	Cocaine, MDMA	$713,564
Solomio	Heroin, weed, cocaine	$232,906
Hippie	Mainly cannabis, a little acid, a little MDMA	$231,711

VikingKing	LSD, psychedelics	$204,803
PantherRed	Speed, cocaine, MDMA, cannabis	$147,450
Thebakerman	Cocaine, MDMA	$140,596

Most vendors do not turn over quite the same volumes. I selected nine medium-size vendors, in terms of numbers of sales. Here, the average monthly turnover is between $10,000 and $20,000 each. Assuming a 100 percent markup on the wholesale price, this means an annual income of between $60,000 and $120,000. A very decent salary, but not exactly drug baron money. This suggests that most vendors are not large-scale international traders, but more likely middle-market and retail dealers. Some sellers are established middlemen who have been involved in the industry for years and have long-standing relationships with importers; they are simply transferring their operations online. But Silk Road has brought new people into the marketplace, ranging from Ace, a twenty-four-year-old who sold "homegrown weed" on the Silk Road in 2012—"I can take about ten to twenty orders a day, so anywhere from seventy to a hundred and forty a week"—to pharmacologists who illegally sell prescription drugs from their surgeries. Angelina was running a medium-size—legal—company when she read about Silk Road in the 2011 *Gawker* article. She set up an operation with a few others, and between 2011 and 2012 completed 10,000 transactions, sourcing wholesale from the producers, and going direct to market. It's lucrative work, but not glamorous, she explained. "We are an importer, manufacturer, and pack-and-ship retailer: it runs like a small internet retailer/

packing, and shipping company," she told *Mashable* magazine in 2012.

DIGITAL REPUTATION

When you buy drugs offline, your choice, to some extent, is limited by geography, and by who you know. On Silk Road 2.0 there is *too much* choice. Thousands of products and hundreds of vendors, operating over several sites. All online marketplaces face this problem. And they all solve it the same way. "Legal e-commerce wouldn't work without user reviews," explains Luke Upchurch from the consumer rights umbrella group Consumers International. "They allow consumers to make more informed decisions about product choice—and allow producers to build up reputations." According to a 2015 YouGov poll, three quarters of Americans check online reviews before purchasing: and 90 percent say they are an important part of the purchase process (though the same number believe businesses write their own positive reviews).

Encryption and the crypto-currency Bitcoin have created the technical conditions that allow Silk Road to exist, but it's the user reviews that make it work. Every drug site has review options, usually a score out of five plus written feedback, and reviewing your purchase accurately and carefully is an obligation for all buyers. "Others need to be able to read this information and predict what exactly the vendor is offering," writes one experienced user to newcomers in a Silk Road 2.0 discussion forum. "When you're going to leave feedback make sure you talk about the shipping times, product quality, product quantity, and the customer service that the vendor provided."

I decided to buy a small amount of marijuana to understand fully how the system works. But this was no tiny corner of the website: there were around 3,000 different varieties advertised, by over two hundred different vendors. I began to scour reviews of different sellers, trying to spot those that others had found to be reliable and trustworthy:

> 1/5: this seller is a fucking scammer, i payed for hashish and now i have 40 grams of fucking paraffin! DON'T BUY FROM THIS CUNT (20 gram of maroc hashish)

No good. After a bit of digging I found one that fit the bill. Drugsheaven was based overseas, but his vendor page advertised "excellent and consistent top quality weed & hash for a fair price." Better still, he had a refund policy and detailed terms and conditions. Drugsheaven sold a staggering array of products: Amnesia Haze, Cheese, White Widow, SoMango, Bubba Kush, Olympia, Messi, Marlboro Gold Stamp, Marlboro 15 Stamp, Gold Seal Afghani, Polo Polm, Lacoste Polm, Ferrari Polm. I carefully scrutinized his reviews. He had a dozen five-star reviews in the last twenty-four hours; and close to 2,000 pieces of feedback over the last four months, averaging around 4.8 out of 5.

> First Order was Lost . . . i got a reship and now im very happy . . . Heaven is One of the best vendors on the road!! Very friendly and very good Communication too. i will be back Soon ;) please Check this vendor . . . 5 Stars

And, importantly, the occasional negative review (a 100 percent record would be unconvincing):

> Product never arrived. Still a trusted vendor though, will order again.
>
> Things happen sometimes, but order at your own risk.

Buyers have reputations to protect too: they are judged according to how much money they have spent on the site and how many refunds they've asked for. A good reputation still matters as a buyer, but for the vendor, your digital reputation is everything. Your name, your size, your promises—all are worthless.

The idea of anonymous marketplaces and reputation systems—how to foster trust in an anonymous world—goes back to the early days of the cypherpunks. But they knew that if everyone was anonymous, then there would be no one to trust. A user could rip someone off under one pseudonym one day, and register another the next. The cypherpunks imagined that people would create long-standing digital pseudonyms: online personas that wouldn't be linked to the "real" you, but would have their own identity and reputation that developed over time. In his 1988 "Crypto-Anarchist Manifesto," cypherpunk cofounder Tim May explained that, in the anarchic digital future, "reputations will be of central importance. Far more important in dealings than even the credit ratings of today." One of the main reasons for the success of Dread Pirate Roberts's Silk Road was the trust he had built up over two years of successful transactions. It didn't matter that no one knew who was behind the mask.

Just like on Amazon or eBay, a positive reputation can take a lot of time to establish. When the original Silk Road was taken down, the vendors lost their old digital profiles, and with them their reputations. "Many of us have spent a lot of time and money just increasing our buyer and/or seller statistics on our respective accounts," complained one seller to the Silk Road 2.0 admins in October 2013. "Is there any chance that any of that data was backed up anywhere and that it can be transferred to the new marketplace?" "No there is no backup," came the flat response. All the old data had gone down with site. "We are all newbies again, I'm afraid." Another administrator added: "It's your job to rebuild that reputation and remind your customers why they chose you in the first place, through exceptional customer service and high-quality products."

With so much money flowing through these sites—and a good reputation the key to getting a slice of it—some vendors try to game the review system. Common tricks include creating fake accounts from which to post positive reviews; writing bad reviews of competitors; paying others to give favorable write-ups; even offering free products in exchange for good feedback. But there is an impressive amount of self-policing here, a genuine drive to identify and remove scammers. The Rumor Mill is the most popular forum on Silk Road 2.0. It is dedicated solely to discussion of vendors and products. In the Rumor Mill threads, reputations and products are aggressively fought over, and scammers are regularly named and shamed. "CapnJack is a scam artist known as KingJoey," writes one disappointed buyer. "He has been ripping people off badly for a long time and he is using this vendor profile to make people think he has really good #4 Heroin left and needs to get rid of it for $170/gram . . . Everyone needs to stay vigilant and avoid this motherfucker." The community will often come together to expose a scam vendor. On one occasion, a group of buyers outed theDrugKing as a vendor writing his own feedback. "I was looking at this vendor a few hours ago, and they had zero feedback. Now they have a bunch, and it's all from users with 6–10 deals," wrote one user. Others forensically analyzed all his posts. "Feedback also has some other things that stick out," said one, "despite a fairly decent job at making the writing look like different people. E.g., 'nice stealth,' 'great stealth,' 'insane stealth,' 'brilliant stealth.'" He was reported to the admins, who banned him. "Well done," wrote one grateful buyer, "for exposing this scammer."

Reputation-based trading produces a powerful but informal consumer-led system of self-regulation, which allows users to make more informed decisions on the products they purchase. When you purchase a drug on the street, you have no reliable way of judging what you're buying, and no recourse if things go wrong. That's why on the streets drug purity is wildly variable: the average purity of street cocaine is 25 percent, but has been found as low as 2 percent—typically cut with mixing substances such as benzocaine by middlemen and pushers. Analysis of seized ecstasy tablets in 2009 found that approximately half had no ecstasy in them at all—rather caffeine and 1-benzylpiperazine. Not knowing what you're putting in your body can have tragic consequences. In 2009–10, for example, a contaminated product led to forty-seven heroin users in Scotland being infected with anthrax. Fourteen died.

Dark net markets provide a radical yet familiar solution to this problem. The user-ranking system provides a safer, systematic, and reliable way of determining the quality and purity of products: trusting the feedback of people who have used them. True, price here is more variable. (In some cases, it is extremely competitive: as of October 2013, cocaine on Silk Road cost an average of $92.20/g compared to an average global street price of

$174.20/g—a saving of 47 percent. On the other hand, its average marijuana price—$12.10/g—was higher than the global average of $9.50/g, and its heroin is particularly expensive, at over twice the U.S. street price.) But according to Steve Rolles of the Transform Drugs Policy Foundation, drug users tend to be willing to pay a slightly higher price if they can be more confident in the product quality.

PURCHASE

I had a market and a vendor. But I didn't have a product. So I accessed the site's internal email system to ask for some advice.

> **Me:** I'm new. Do you think I could just buy a tiny amount of marijuana?

Twelve hours later, I received a response:

> **Vendor:** Hi there! My advice is that starting small is the smart thing to do, so no problem if you want to start with 1 gram. I would too if I were you. I hope we can do some business! Kind regards.

How polite! I follow his advice and opt for the smallest amount: 1 gram of marijuana. It costs 0.03 Bitcoins, which comes to around £8 (plus free shipping and delivery). I click "add to cart." Now all that's left to do is pay.

PROCEED TO CHECKOUT

Markets adapt to problems. Every crisis is swiftly followed by innovation. The dark net markets are getting smarter all the time.

The most important conundrum here is how to make sure buyers don't get ripped off. When you buy something from reputable websites you part with your money before you receive the product, even though you don't know the seller. This is because you trust that the system works. Sites like Amazon have tried-and-tested mechanisms to assess products and deal with disputes, and are governed by legislation to cover consumer rights and uphold trading standards.

Upfront payment was also the preferred method on early drug markets like the Farmer's Market. Buyers would place an order, hand over their money, and wait, hopefully, for a package to arrive. But it is a model that is too unreliable in drug markets. Silk Road solved this problem by adopting something called an "escrow."

Every dark net market operates its own internal wallet system. On the original Silk Road, a buyer would create a site-specific wallet, into which they transferred Bitcoins from another wallet—typically one held on their own computer. Once the order was placed, the buyer would transfer the correct amount of Bitcoins from their Silk Road wallet to an escrow—a wallet controlled by a Silk Road administrator. The vendor would be notified that the money had been placed in the escrow and send the order. When the buyer received the product, they notified the site, who released the money to the vendor's own Silk Road wallet. On arrival to the original Silk Road, a message from DPR advised all newcomers: "Always use the escrow system! This can't be stressed enough. 99 percent of scams are from people who set up fake vendor accounts and ask buyers to pay them directly or release payment before their order arrives" (this is called the Finalize Early (FE) scam). Escrow systems go back centuries. The word escrow derives from the Old French word "escroue," meaning a scrap of paper or a roll of parchment. eBay has an optional fee-based escrow system. But it was unheard of in the drug market, where the concept of consumer protection was nonexistent.

But escrow is flawed, because you still have to trust a drug website with your Bitcoins. As the six months of mayhem following the fall of Silk Road showed, sites are liable to theft by site admins, attacks by hackers or confiscation by the police. In all cases, you have little recourse. When in February 2014 Silk Road 2.0's entire cache—$2.5 million worth of Bitcoin in buyers' and vendors' Silk Road wallets—was stolen, Defcon took to the forum to announce the bad news. Apologizing profusely, he added that "Silk Road will never again be a centralized escrow storage . . . I am now fully convinced that no hosted escrow service is safe." Defcon proposed a new, even more secure, payment method called multi-signature escrow. This way, after a purchase is accepted by a vendor, a new Bitcoin storage wallet is made. The vendor approves on order, the buyer approves on receipt, then the site approves (or declines if there is a problem). But the money is released only when two of the three sign off on it with their PGP keys. No one party can disappear with the money. It's like a safe where all the key holders must be present to unlock it. Any problems, and the buyer gets his money back.

Some of the newer markets have started using "multi-sig" escrow. The feeling in the dark net markets community is that no site without it can be fully trusted. One user summarized the mood in a March 2014 post on a Reddit forum for dark net markets, entitled "Do. Not. Use. Or. Create. New. Sites. Without. Multi. Signature. Escrow":

> Okay guys, time for some tough love. Remember that Simpsons episode where Homer walks into the same wall five times and says ouch? That's you signing up for another site without Multi-signature Escrow (Multisig) . . . So fuck new sites. Seriously, fuck them. DO NOT sign up for sites because they (1) have pretty pirate graphics (2) claim to be competent (3) claim to be different (4) claim to be charitable (5) have a catchy name (6) Claim to be the fifth reincarnation of the Dalai Lama (7) use any other gimmick, branding or marketing technique. ONLY sign up for sites that claim to use proven security techniques, such as Multi-signature escrow. If you get scammed, you are an asshole because you have proven to them that it is profitable.

Defcon agreed. "[Multi-sig is] the only way this community will be protected long-term. I am aggressively tasking our dev[eloper]s on building multi-sig support for commonly used Bitcoin clients. Expect a generous bounty if you have the skill to implement this." When trying to choose between different untrusted markets, multi-sig makes perfect sense. It's a strike against a centralized system. It provides an extra layer of consumer protection.

Dark net markets are always adapting to the challenges they face. Bitcoins aren't quite as anonymous as some people think. Because of the way the technology works, every transaction needs to be recorded in the public blockchain, to prevent double spending. So if I sent my Bitcoin directly from my wallet to my Silk Road wallet, the blockchain would keep a record of it. My privacy is intact, because no one knows that my Bitcoin wallet belongs to me. However, researchers have found that with careful statistical analysis of transaction data, some transactions can be de-anonymized. And if you transfer money in and out of your Bitcoin wallet using a real-world bank account, which most people do—including me—your Bitcoin transactions aren't anonymous at all. They lead to you as directly as paying with a credit card.

Developers therefore created "tumbling" services. I send my Bitcoins to a central holding wallet that pools together several users' transactions at once, mixes them together, and then sends them along to their final destination. My Silk Road wallet now has the correct amount of Bitcoins—but they are different than the ones I sent, and so cannot be traced back to me. This is extremely clever, and amounts simply to a small-scale micro-laundering system. But it suffers the same weakness of the centralized escrow systems: you still have to trust someone else with your coins. What's more, the tumbling service charges a small fee for its use. So other developers are working on free, open mixing services. CoinJoin, for example, works in a similar way, except that there is no central holder of the money: a number of anonymous users place their payments into a temporary address, which mixes them up and sends them only when everyone has approved the transactions.

It's about to get smarter still. Many in this community are aiming for something bigger. When I started this chapter my aim was to understand how this community had created a marketplace that people can trust in such unlikely circumstances. It turns out their ultimate goal is the precise opposite. The escrow payment system, multi-signature, and CoinJoin are all part of a bigger plan to create a market where you don't need trust, because everything is guaranteed to work with powerful encryption and decentralized systems that can't be shut down or censored. They want to create a trust-*less* market. The future of these markets is not centralized sites like Silk Road 2.0, but sites where everything is decentralized, where listings, messaging, payment, and feedback are all separated, controlled by no central party. A site that would be impossible to censor, or to close. Of course, law enforcement agencies can't allow these sites to operate with impunity. Doubtless they are working out how to either knock them offline, or, more likely, undermine the confidence people have in them. What's more, it's hard to imagine that traditional drug dealers will sit aside as their business model is torn up. Perhaps they, too, will shift their operations online: according to the *Guardian*'s 2014 British Drugs Survey, 45 percent of people who have sold drugs for profit have purchased them from the internet. But there is no shortage of clever, motivated individuals committed to keeping these markets up, running, and improving. For many of them this is more than about drugs. It's the digital front in a war for individual liberty.

I follow the last steps: transferring my clean Bitcoins to the vendor's wallet, and click "pay." I'm immediately taken to a new page where I find a short message: "Your transaction was successfully completed."

STEALTH

But there is still one final hurdle to overcome: actually getting hold of my drugs. For all the clever payment and tumbling systems, I have to provide a real-world address in order to receive my product. Some people use what's called a "drop address"—an abandoned house with a functioning mail box. The majority, me included, simply supply their home address, trusting in the power of "stealth." Vendors are often ranked according to how quickly and easily their packages are delivered—or how cleverly they disguise the product. Although my vendor's stealth methods were not discussed in his reviews—for fear of tipping off the authorities, I learned in one forum—they were highly praised. And quite right too.

One morning, five days after I've placed my order, a white package arrives at my house. It's about the size of a postcard, but a little bulky—padded with bubblewrap. The name and address I'd entered into the site using my PGP key was printed on to a small sticker. It looked, felt, and smelled exactly like every other item of post I'd received that week. Inside, the product was carefully sealed, the correct weight and, according to an expert friend of mine, appeared to be extremely good quality.

The last thing I had to do before closing down my account was to leave a short and simple review: "The drugs arrived as described. 4/5."

MARKET PRESSURE

Buying drugs on Silk Road is extremely pleasant. Browsing through the endless options, I was bombarded with special offers, free packaging, complimentary extras. Vendors were attentive and responsive. The products were of a (reportedly) high quality and competitively priced, according to my research. Here, the customer is king.

The drugs market has always been characterized by local monopolies and cartels. Dark net markets have introduced a new dynamic to this world, what the famous post-war economist Albert Hirschman calls "Exit" and "Voice"—two features that keep organizations working to the benefit of those using them. Unhappy buyers can now express their "voice" via feedback; and can "exit" a poor vendor in favor of one of over 800 others. That means vendors are forced to compete for buyers, and are pinned by the review system. Through the introduction of clever payment mechanisms, feedback systems, and the injection of real competition, power has shifted away from sellers and back to consumers. There is no clearer indication of who rules these markets than one of the last posts on the Silk Road 2.0 forum by the hard-headed administrator Libertas in November 2013:

> Hi all. My apologies to all you experiencing slow Customer Support response times . . . We are implementing changes to ensure that messages cannot be missed in future, and again, I apologize for any inconvenience that any delays in responding to your tickets may have caused. Libertas.

Exit and voice on dark net markets are doing precisely what economics textbooks predict: creating a better deal for consumers. The most surprising statistics of all in the vendor data I presented earlier in the chapter are not the volume or range of drugs on offer, but the satisfaction scores. Over 95 percent of the 120,000 reviews score the maximum five out of five; only 2 percent were one out of five. When Professor Nicolas Christin analyzed feedback scores on Silk Road in 2012, it was almost identical.Once this sort of consumer-focused market is unleashed, it infects everything. The Silk Road 2.0, Agora, Pandora Market, and the others are already competing with each other on security, commission rates, and usability. In April 2014, Grams, a search engine for the drugs market, was launched. Grams searches the largest markets for products, making it easier to find what you're after. Grams has recently introduced trending searches; and vendors can even buy sponsored keyword listings space for their sites and products. According to James Martin, author of *Drugs on the Dark Net*, some vendors are even beginning to brand their opium or cocaine as "fair trade," "organic," or sourced from conflict free zones. "We are a team of libertarian cocaine dealers," writes one dealer. "We never buy coke from cartels! We never buy coke from police! We help farmers from Peru, Bolivia, and some chemistry students in Brazil, Paraguay, and Argentina. We do fair trade!"

In November 2014, yet another major raid (codenamed "Operation Onymous") led by Interpol and the FBI took place, and Silk Road 2.0 was shut down, with a number of its chief operators arrested. At the time of writing, the specific details of how law enforcement did it are unclear. Yet there remain at least twenty or so of these markets currently in operation, selling far more drugs than when I started writing this chapter in August 2013. They emerged from the turbulence of late 2013 more secure and customer-friendly, armed with multi-sig escrow, CoinJoin, and search engines. Individual markets or vendors will, of course, be caught again in future. But each time the hydra grows a new head. Markets learn. This all empowers the consumer, forcing the quality of the drugs and the sites to improve still further.

So it won't be Silk Road 2.0, or perhaps not even Tor Hidden Services, that transforms the drugs industry. But now consumers are in charge: it will never be the same again.

What this means for the war on drugs is not clear. In October

2013, an ambitious study brought together data from seven different drugs-surveillance systems around the world. The paper, conducted by the International Centre for Science in Drug Policy, concluded that the war on drugs is failing. Illicit substances are more available than ever. Since President Nixon declared war on drugs in 1971 there has been no noticeable curb in supply, and certainly not in demand. Dark net markets make drugs available more easily. That's nothing to celebrate. It will tend towards higher levels of use, and drugs use—legal or illegal—creates misery. There is violence and corruption at every point in the supply chain as drugs move from producers to street dealers.* The longer the chain, the more violence and suffering; the more profits for dealers; and the more substances are cut and mixed. Dark net markets won't fix all of this—but they will reduce the length of the supply chain. Although reliable figures for the exact cost of street trading are notoriously difficult to find, according to The Drugs Policy Alliance, in 2013 alone the U.S. government spent $50 billion a year on the war on drugs, yet still arrested 1.5 million people on non-violent drug charges; and around 44 thousand people died from an overdose. A study by the United Nations suggests that the costs associated with drug-related crime (fraud, burglary, robbery, and shoplifting) in England and Wales were equivalent to 1.6 percent of GDP, or 90 percent of all the economic and social costs related to drug abuse. This is also likely to reduce, as buyers cut out the street dealer. History suggests that those who want drugs will always find a way to get them. On Silk Road, they can

* One user of the website told me that he expected street dealers to begin buying products in bulk to break down and resell.

get a better product and with fewer negative risks associated with buying drugs on the street.

Online drugs markets are transforming the dirty business of buying drugs into a simple transaction between empowered consumers and responsive vendors. It's not anonymity, Bitcoins, or encryption that ensure the future success of dark net markets. The real secret of dark net markets is good customer service.

6

LIGHTS, WEB-CAMERA, ACTION

I recognize Vex from across the street. She looks just like she does on my computer: early twenties, elfin, attractive, slim, with two nose piercings. She bursts into the café where we've agreed to meet, flustered because she's late. She's just finished a "cam-show," where she earnt more money in an hour than many people earn in a week. Vex is one of the world's top cam-models. She makes her living by posing, chatting, stripping, and masturbating live to thousands of people each week from the comfort of her bedroom. Every day or two Vex will perform a show on the website Chaturbate to between 500 and 1,000 viewers from all over the world. But there is a lot more to her job than taking off her clothes. In a typical day, she spends a number of hours checking and updating the several social network accounts and websites she maintains as part of her impressive brand. She communicates with fans, responds to emails and posts messages about her upcoming shows. She sends thank-you notes and, occasionally, gifts to regular viewers and big tippers, and takes photographs and videos of herself for members of her fan club. It is a full-time job. But Vex considers herself an amateur, an "ordinary," one who enjoys performing online, and one who is lucky enough to be able to make a living from it. She's not alone.

The net has made pornography extremely easy to find. But it has also made it very easy to make and sell. With the advent of good quality webcams and broadband it is now possible for anyone to produce pornography from their home. We ordinaries are flooding the net with home-made pornography—everything from sexy selfies to hard-core videos. It's no longer professional porn stars that dominate the market: it's us. The four most viewed pornographic websites today are all free, and all feature primarily amateur material. The fifteen most visited porn sites contain between them almost two million explicit amateur videos. Although precise figures are difficult to obtain, the traditional porn industry—although not dead by any means—has been hit hard by these trends. The Free Speech Coalition estimated global (and American) porn revenues have shrunk by 50 percent between 2007 and 2011, due to the amount of free pornography available online.

Long before the first webcam was connected to the internet in November 1993 people were using networked computing to get their sexual kicks. In the 1980s, users of Bulletin Board Systems would scour forums for a member of the opposite sex to flirt with and, hopefully, have "compu-sex" with. Erotic story groups where people could read and write steamy prose became extremely popular on Usenet in the 1990s. The first erotic stories Usenet group, rec.arts. erotica, was created in May 1991, which was quickly followed by a number of spin-offs and subgroups catering for every predilection, including one titled alt.startrek.creative.erotica .moderated.

Although statistics about how much pornography is on the net are usually marked by exaggeration and moral panic, there has certainly always been a lot of it about. One pornographic Bulletin Board System earned 3.2 million dollars in 1993 by selling hard-core images and videos to thousands of subscribers. By 1997, there was somewhere between 28,000 and 72,000 porn sites online. Porn is now estimated to comprise between 4 and 30 percent of all websites.

In April 1996, an American university student named Jennifer Ringley registered a website named "JenniCam." She filmed herself performing a range of activities—from brushing her teeth to stripteases—and streamed it live on the site. Jennifer was the very first cam-model, one whose motivation, she claimed, was to give viewers "an insight into a virtual human zoo." At its peak, four million people were watching the jolting and stuttering frames of a life lived in front of the computer. Ringley quickly realized the financial possibilities her popularity presented. In 1998, she divided her site into free and paying, with a fifteen-dollar annual fee allowing access to new pictures every two minutes (if you didn't pay, you'd have to wait twenty minutes).

Within months, thousands of aspiring performers had created copycat sites across the web. The majority were set up by amateurs, and more often than not were pay-to-view. They included several spoof JenniCam sites, and even a "Voyeur Dorm" cam, featuring five college students leading normal lives in front of thirty-one cameras running around the clock. In 2001, Live Jasmin, an erotic internet-based reality-television show, was set up by the Hungarian entrepreneur György Gattyán. The site soon morphed into a place for amateur and aspiring models to go to perform to paying viewers. Camming was becoming a gainful means of employment, albeit still a very niche one. Then a new generation of camsites arrived: MyFreeCams in 2004 and Chaturbate in 2011 offered free, professionally produced and regularly maintained websites. Camming experienced a surge in popularity.

Chaturbate is one of the largest camsites in terms of unique site visits, second only to Live Jasmin. At any one time, there are around 600 models online, from all over the world, sitting at home with their webcams on, and performing live for people who have logged in and joined their Chaturbate "room."

There are no signing-on fees or subscriptions on Chaturbate, and the performers make their money from "tips" from viewers. Cam-models will often perform once or twice a day, and a typical show might last an hour. During a show, viewers can tip the models in Chaturbate tokens, which can be bought on the site. Shirley, the glamorous, American, thirty-year-old chief technical officer at Chaturbate (who seems to more or less run the site single-handedly), explains to me how it works. Cam-models use Chaturbate a bit like a franchise. They have to pay 60 percent of any tokens they earn to the company, who offer the server space, web maintenance, payment processes and, as Shirley puts it, "the power of the Chaturbate brand." It is a powerful brand. Chaturbate receives approximately three million unique visitors a day. That's a lot of potential tippers.

You don't need to look like a porn star to become a performer on Chaturbate, and most do not. A 2013 study of 7,000 professional female porn stars from the United States revealed that the average actress is 5 foot 5 inches tall with measurements of 34–24–34. There are certainly cam-models that look like that too, but also plenty who don't. Today there are probably around 50,000 or so cam-models working from bedrooms and studios around the world, mostly from the United States, Canada, and Europe. Although men, transsexuals, and couples also perform, typical cam-models are women aged between twenty and thirty. Beyond that it's hard to be exact. Some do it full-time, like Vex. Others dip in for fun, or to supplement their income. On Chaturbate you will likely find (and I have found): middle-aged couples going at it hammer and tongs, naked men playing the guitar, bored-looking females sitting silently, a transsexual orgy, a penis being vigorously masturbated (of course), as well as countless other performers in various pre- and post-coital states. The most popular models can easily have more than 1,000 viewers any time they perform, but some will only have a few dozen watching. There is something for everybody here.

According to *The New York Times*, camming has become an enormous, global industry that generates over a billion dollars a year, which is about 20 percent of the pornography industry as a whole. There is even a large and growing cam community, with several online forums and support groups where cam-models can meet, talk, and exchange ideas. Camming is a vibrant and flourishing world, and one almost entirely made up of "ordinaries" with webcams, just like Vex. But what is it that makes ordinaries so popular?

Vex hadn't planned to become a cam-model. She had signed up to pose for a series of nude photographs for an alternative soft-core company called God's Girls to earn some extra cash while she was at university. One day, she overheard some friends at God's Girls discussing camming as a good way to make money. After a spot of research, she bought a webcam, and signed up to perform private shows on Skype with a local camming company. "I was very nervous the first time I did it. I talked way too much," she tells me. "I had twenty visitors in my room, and that felt insane! I think I made about thirty pounds." She then joined Chaturbate, and almost immediately started making enough money for it to become her sole source of employment.

I ask Vex why she thinks she is so popular. "Traditional porn tends to be standardized and unrealistic," she replies. "I guess I'm a real person in a real room." This view is one put forward by Feona Attwood, professor of cultural studies at Middlesex University: "It's a better kind of porn: somehow more real, raw, and innovative than the products of the mainstream porn industry."

"Are you more 'real' than mainstream porn?" I ask.

"Well, the best way to find out," Vex says, "is to come along to one of my shows."

A RATE-ME INTERLUDE

Over the last decade, our lives have been turned inside out. The amount of information we now routinely share about ourselves is staggering. Around the world there are between 1.73 billion and 1.86 billion active social media users across all platforms: posting status updates, uploading videos, describing or sharing photos of where they are, what they like, who they are with. On Facebook alone, users upload somewhere between 20 and 140 billion items of content every month. Some psychologists think that social media is popular because it taps into a hard-wired need to seek affirmation from our peers, "a deep evolutionary need for community," and a very natural desire for enhanced reputation. Sharing our every intimacy is a shortcut to fulfilling innate needs that are present in all of us: needs for affection, relationships, belonging, self-esteem, and social recognition.

Social networking sites have developed clever new ways to encourage you to share increasing amounts of information about yourself. This is because personal data is an extremely valuable commodity. The more you provide, the more of your digital breadcrumbs can be collected and sold on to companies who want to sell you something. There is a multi-billion-dollar industry that does nothing except buy and sell the personal data we share online.

The MIT professor Sherry Turkle thinks that the fact we're always logged on, and always being observed, is turning many of us into personal brand managers, who carefully sculpt our online identities and obsess over what others think of us. In her 2011 book *Alone Together*, Turkle interviews hundreds of children and adults, and documents the new instabilities in how we understand privacy and personal identity online. She records young people spending hours a week carefully selecting their Facebook likes, and photoshopping mildly unflattering photos: "What kind of social life should I say I have?" they ask. She calls this "presentation anxiety."

Young people today have different views about privacy compared to their parents. Pew Research's 2013 study of American teens, social media, and privacy found that young people are sharing more information about themselves than they did in the past. In 2012, 91 percent posted a photo of themselves, (up from 79 percent in 2006); 71 percent post their school name and the city or town where they live; half post their email address. That's not to say young people don't care about protecting their privacy online—surveys show they do—but rather that they regard their privacy in terms of retaining control over what they share publicly, rather than limiting what they share.

This sharing epidemic stretches into the intimate. According to a 2014 study by the Security Firm McAfee, around half of U.S. adults have sent sexy photos or texts of themselves via their smart phones. A similar poll conducted by YouGov found that approximately one fifth of British adults under forty have engaged in sexual activity in front of a camera; 15 percent have appeared naked in front of a web-cam. Of those who said that they had taken "selfies," 25 percent also admitted to taking "sexy selfies."

Sharing images of ourselves or our body parts is not a recent phenomenon. A lengthy history could be written about penises alone, beginning perhaps with the 28,000-year-old image of a phallus in the Hohle Fels cave in Germany. But the instantaneous connectivity of life online has certainly enabled us to do so more easily, and to a wider audience.

The website 4chan (which you'll remember from chapter 1) has a socializing board called /soc/, which is a space specifically for cam-models, meet-up groups, and extremely popular "rate-me" threads. It's a sort of ground zero of exhibitionism. A rate-me thread is exactly what it sounds like. Every few minutes on one of the hundreds of threads on 4chan, someone posts a photo (often naked), accompanied by a message inviting feedback. Viewers respond—sometimes positively, sometimes not, and almost always with a score out of ten.

Penis rate-me threads are especially popular on 4chan, and

have several subgenres: small penis threads, large penis threads, skinny penis threads. "Rate my dick please," wrote a user in one thread, next to a picture of his member, "and feel free to share yours." Within a minute, he'd received the following comments:

> Thick, long 8/10
>
> Veiny slightly weird color 5/10
>
> Fucking huge! 10/10
>
> I'm not even gay and I'd suck it. Absolutely jealous.
>
> 9/10
>
> 9/10 would swallow

Others took up his offer and shared theirs, often pictured alongside household objects to provide a handy scale: a television remote control, a roll of toilet paper, a small bottle of orange juice.

One regular poster is Joe, a twenty-year-old office worker from London. He posts photographs of himself in the "skinny guy" threads, where slim males share pictures of their bodies. "I post for the attention, I suppose," he tells me. "I'm fed up with people always mentioning how thin I am. It's refreshing to hear that somebody appreciates my body. Whether they're male or female isn't a concern. I'm only listening to the compliments: it's a confidence boost." Joe has saved every positive comment he has received—there are hundreds—in a folder on his desktop.

I click through countless rate-me sites on 4chan and across the web; sites dedicated to babies, puppies, hairstyles, as well as to arms, muscles, and poo (believe it or not). Every one contains thousands of images, each with its own list of comments and ratings.

The rate-me phenomenon spreads much further than specific

websites. In 2011, a Facebook group titled "hottest and cutest teenagers" was created. It spawned a number of similar groups or forums on the site, with countless teenagers participating. The pages provoked an outcry from concerned parents and security experts, and some of the sites were quickly shut down. But each time one was closed, another appeared in its place: "the Most Beautiful Teen in the world," for example, in 2012: "cutest teens" in 2013. At the moment of writing, there are at least twenty-five of these groups or events on Facebook. Rate-me videos are also proliferating. Over the last few years thousands of young teenagers have uploaded clips of themselves on to YouTube with the caption: *Am I pretty or ugly?*

In whatever form it takes, it is clear that more people are sharing more personal images about themselves than ever before. Vex is not an outlier. She is the visible tip a very large iceberg.

LIGHTS, WEB-CAMERA . . .

Three weeks later I'm at Vex's house in the north of England. Vex is hosting a special show today. Like most cam-models, she usually performs alone. But this evening two other cam-models will be performing with her. It's a "girl-on-girl-on-girl" show. Vex does one of these every few months, she tells me. They are always well publicized and extremely popular. A big audience is expected.

Vex lives on a polite redbrick road. Her four-story house is spacious but cluttered with pieces of art and retro furniture. She gives me a quick tour and takes me up to a fairly small, messy bedroom at the top of the house.

Two girls are sitting on the bed. They're both in their early twenties, like Vex. Auryn is from Canada. She's tall, black, slim, and heavily tattooed. She's been camming full-time for about a year, she says. Blath is shorter, white, has striking pixie-features, and hair that's dyed a greeny-blue color. She's studying photography, and cams part-time for a bit of extra cash. The girls all worked together at God's Girls, and have been friends ever since.

The show is due to start in an hour, and the girls hurry to get ready: cleaning, arranging furniture, setting up the lighting. One fan has sent Vex a bottle of champagne for the evening, which she produces, pouring glasses for the girls. "What shall we wear?" Blath asks. "Neutral colors, I think," Auryn replies. "Matching?" "Yes." They start discussing tax returns while mixing sparkle glitter into a bottle of baby oil. Vex walks over to a small chest of drawers. "Right: here are my dildos," she explains. "The sex toys have been sterilized, of course," she adds, handing me one. "Ah. And wet-wipes. A camgirl's staple. We have to put an unusual amount of objects in our vaginas, you know."

We are already thirty minutes late, and Vex, Blath, and Auryn are just about ready, dressed in skimpy tops, long socks, and frilly knickers. Before we go live, Vex explains, Auryn and I need to be "age-verified" by Chaturbate. She emails photographs of our driving licences to the site's moderators. Chaturbate has a very strict policy regarding age, for obvious reasons, but I'm not entirely sure why I've been included. "Don't worry," Vex says. "It's just in case you accidentally walk past the screen or something."

Cam-shows need to be carefully orchestrated. Vex predicts the show might last for at least two hours. It's a lot of time to fill. The girls first agree on what the boundaries will be tonight. It's decided they will do "pussy play," close-up shots, and use vibrators, but not dildos. Then they need to decide on the room incentives. There are a lot of clever ways to make use of Chaturbate's tipping system. Most cam-models set staggered targets: an escalating number of tokens for increasingly explicit acts. Some have a set tip menu that doesn't change. Usually there is a single overall "token goal," which the performer hopes to reach in the course of the show, and which will result in a show's finale. Vex, Blath, and Auryn decide on something a bit more complicated. They set up two "Keno" boards, a tipping system that works a little bit like bingo. Vex creates eighty numbered boxes, each with a token price. You open each box by paying the amount, but only one in four of the boxes contain prizes. The first Keno board contains only "soft-core" rewards; the second, more expensive board, only "hard-core" acts. All three shout out possible prizes, and Vex types them in on her Chaturbate account:

Blath and Vex making out

Blath and Auryn making out

Auryn and Vex making out

Auryn, Vex, and Blath making out

Boobies

Oiling up

Nipple kissing

Pussy flash

Extra-gropey make-out between two girls of your choice

Oral triangle

Vibrator: one minute

Vex adds a “firm handshake” for fun. “Gosh,” she says, “there are a lot of prizes.” They decide that the final goal will be a password-protected “cum-show.” Only viewers who have paid 200 tokens each—approximately twenty dollars—will be able to watch.

Finally we are ready to go. The three of them are sitting on the bed in sexy clothes, arms around each other like a school gang. I am sitting just off-camera, two feet away from the bed, a pad of paper and my laptop on my knees, logged into Vex’s virtual room. It all feels a little strange, to say the least.

One last swig of champagne. “Right. Are we ready?” asks Vex. Loud cries of yes all around. “Okay, let’s go!” She turns on her camera, and suddenly the three of them appear onscreen in Vex’s digital room.

. . . AND ACTION

As soon as Vex’s camera is turned on, Chaturbate automatically sends a notification to her 60,000 Chaturbate fans, informing them that she is now in her room and ready to perform. We are live. If I expected a dramatic transformation when the cam starts rolling, I was mistaken. The show begins in a quiet and pleasantly haphazard way. The first ten minutes are characterized by minor technical glitches. The frame speed is too slow. The Keno boards seem to be deleting prizes. The computer, which is positioned on a table three feet away from the bed with an expensive clip-on webcam, is still being adjusted by Vex as people start to join her virtual room.

And *a lot* of people are joining. The first to arrive is a regular visitor from the United States called Danny. “Hey, Danny!” the girls say in unison. "You're the first one in!" More regulars join and start communicating via the chat box. When you watch a show on Chaturbate, your screen is split in two. On the left side is the live video stream from the webcam. On the right is a constantly updating chat box, which is where the audience communicates to the performers and each other. Vex, it quickly transpires, has a very large fan base who watch nearly all of her shows. As the regulars join, they typically greet the performers by typing in the box, and the girls greet them back by talking into the camera. Oh hey, Night-shadow, how are you? Welcome back, Stroker! Ox! So nice to see you. But far too many people are pouring in to greet each individually. Within five minutes, over 600 have joined us.

Chaturbate has a handy—if slightly ruthless—trick to help the performers manage these large crowds. Visitors with no tokens in their account appear as gray in the chat box. Everyone calls them "grays." These are the freeloaders. Performers sometimes decide to "turn the grays off," which means they can watch, but not talk in the chat box. Users whose names appear as light blue have tokens but haven't tipped more than fifty tokens in the last twenty-four hours. They are the cautious, parsimonious tippers. Dark blue means they have plenty of tokens, and they spend them regularly. If a dark blue says something in the chat box, they are more likely to get a response. It's a board-specific class system.

Within thirty seconds, the first tip arrives. It's ten tokens from Danny. Every time someone tips, it appears next to their name in the chat box, and there is a small "ding" noise. Ding! Someone throws in enough to open a Keno prize: fifty-three tokens. "Thank you, Bumbler!" says Vex. It's the Firm Handshake. The three girls laugh, and earnestly perform their task. I laugh. The 2,000 or so people now in the room laugh too, albeit in text speak.

This show isn't all about sex. It's about entertainment and socializing. The girls seem to be having a lot of fun. They draw on each other's bodies. Poke fun at each other. Clothes come off, and then go back on again. Vex falls off the bed. Auryn chats with viewers about various news items. There is a *Ding!* every three or four minutes, and every five minutes or so a prize is won and performed. After the handshake comes the "extra-gropey make-out between two girls of your choice." "Oh, that's a good one!" says Vex. And for the next two minutes, she and Auryn perform what must be an extra-gropey make-out.

The chatter is constant. Most conversations are led by a relatively small band of regular fans, who all know Vex and appear to know each other. I identify around two dozen regulars, and begin to contact them individually. They are a fun and welcoming bunch. One, Bob, is a thirty-two-year-old single man from the UK and "not a hermit who lives in a virtual world!" he tells me. He has made a lot of friends in and around Vex's room: "I have come to consider the time spent hanging out in cam rooms as important as spending time with what I would call 'real-world' friends," he says. There is certainly a social aspect to it. This show lasts for three hours, and the majority of viewers stay from beginning to end.

I can see why. Vex is extremely friendly, funny, and attentive, and a natural performer. She doesn't try to be perfect. She tries to be normal. Danny tells me that the best camgirls are the ones who can laugh at themselves: "Vex does something silly on at least a weekly basis," he explains, "if not more regularly. So aside from being breathtakingly attractive, it's always interesting to see what games and incentives she has planned." Before meeting Vex, I watched several of her shows from the other side of the screen. It looked effortless. But being so close to the action I get a sense of how hard it must be. Vex—with excellent support from Blath and Auryn—is keeping thousands of people entertained with nothing but herself and a webcam. This takes a lot of imagination, according to Danny—taking your clothes off and touching yourself gets stale and predictable very quickly.

In a world awash with hard-core pornography, it's personality and inventiveness that set the best apart. Aella, another of the more popular camgirls on Chaturbate from the United States, often puts on entire shows as a mime. She humps chairs, plays with toy gnomes, and sometimes does shows where she doesn't take any clothes off at all. Even so, there is a lot of time to fill. On WeCamGirls—a community website for cam-models—the most popular forum is for sharing lesson plans, tricks, tips, and ideas. "You have to be really, really imaginative," Vex agrees. "It's not easy."

Vex has a strange knack of being able to twist everything into a part of her show. Her cat, Duchamp, keeps wandering in and out of the room, and is often brought on camera.* Last year she started a sticker club, and created an enormous wall chart, with a list of regular viewers' names. If you see Duchamp—sticker. If you tip a certain amount—sticker. It was wildly popular.

---

* Cats wandering in and out of amateur pornography has become an internet meme all of its own. There is an excellent blog dedicated to "Indifferent Cats in Amateur Porn." Duchamp's sister Liesl has made an appearance there.

Although I was well hidden, I worried that I would ruin the show—perhaps I'd put off the girls, or the viewers. That was surprisingly not the case.

"So, we have a writer here today with us, guys," says Vex to the camera suddenly. I start shaking my head. Violently. "And if you tip one thousand tokens in the next minute, he's promised he'll say hello." *Ding! Ding! Ding!* Vex jumps off the bed and drags me over.

"Hello, everyone," I say.

"Hello," reply a dozen light and dark blue names.

"I'm writing a book too?" types one.

"I'll pay you ten thousand tokens to suck his cock," adds another.

"Guys! Stop it! What shall his Chaturbate name be?"

"The Pen-is-Mightier!"

I hastily retreat. Despite my appearance, the show is a success. The tips are coming in rapidly. That's perhaps because Vex's real talent—refined over hundreds of shows—is to keep business ticking over. "Remember, guys," she says after spanking Auryn with a horse crop, "there will be a very special password-only show after this, and it's only two hundred tokens to join."

THE TIP-BOMB, AND OTHER FINANCIAL MATTERS

About forty minutes into the show, something remarkable happens. One user pays a big tip—999 tokens—for no apparent reason. Someone else matches him. And then so does another. The girls freeze. For about three minutes there is a tipping frenzy. When it's over Blath, Vex, and Auryn are each about £60 richer, and all of the hard-core Keno goals have been met. This is known in the trade as a "tip-bomb." They now owe the room a Keno board's worth of seriously heavy petting.

"Holy fuck," says Vex to the fans. "I have *never* seen a board go so quickly. We have so much shit to do. Awesome!"

"What the hell are the prizes, again?" says Auryn.

"Someone needs to go down on me," says Blath.

"Shouldn't we just, like, do everything to each other?" suggests Vex.

The tokens continue to flood in but I am beginning to notice that the tipping largely follows its own rhythm, unconnected to what's happening onscreen. There are lists and leaderboards everywhere on Chaturbate. The top tipper in today's room will win a prize: custom Polaroids that the girls will make after the show. Every few minutes a tipper leaderboard is displayed in the chat box (titled "most *valued* tippers," and not "highest tippers"). When you first join Chaturbate there's a leaderboard of popular rooms, and a list of the most popular performers. Vex displays a list of her top all-time tippers on her profile page. On MyFreeCams, cammodels have a "camscore," which is a formula based on average tokens received per minute.

This is an extremely clever system, applying a soft competitive pressure on everyone to keep spending. Vex's regular tippers are loyal to her—and on the whole they tip because they appreciate the show she puts on, or simply because they like to see her happy. But some float around different rooms, dispensing enormous tips to several girls, just to be top of their list. (Some take being a model's highest tipper very seriously indeed, and do not like it if someone dislodges them as The Favorite.) I suspect the tippers here have one eye on each other, too. Tipping big certainly impresses the 5,000 others in the room. If you tip-bomb the Keno board, everyone gets to enjoy the show. Some of Vex's fans really do tip her very well indeed. One tells me via email that he typically spends $500 to $700 every month on cam-models. Another viewer tipped $800 in this show alone.

These fans make Vex one of Chaturbate's top earners. After a bit of head counting, she estimates her salary to be around £40,000 per year—that's about $60,000. The most she's ever earned in a single show is about £1,000 ($1,500) thanks to enormous, unsolicited tips, often from regulars. But there are other girls who make even more. Vex told me of one girl who'd made $20,000 in a single month. The money from tonight's show will arrive in a couple of weeks. The tokens she earns will get exchanged into dollars, Chaturbate takes its cut, and transfers the rest on to a prepaid card called a Payoneer. Payoneer, in turn, takes a small percentage every time it is used. Vex, Auryn, and Blath all complain about this. "There has to be a better way," Vex says after the show. "Have you heard of this new thing called Bitcoin?"

Tokens are not the only perk of the job. Vex, like many performers, has a "wish-list"—a personalized page on Amazon filled with things fans can buy her. Chaturbate's chief technical officer Shirley tells me one girl had a fan pay for breast enlargement; another was bought a washing machine. Vex said she was once sent a set of Le Creuset cooking pots. I found one cam-model wish-list that included books of left-wing political and social criticism, a Black & Decker Dust Buster, and something called a Ruckus motorcycle atv car valve spring compressor tool.

Partly because of the possible rewards, Vex tells me that cam-modelling sites are becoming increasingly crowded. There are more performers joining every week. More girls means more competition, and this is driving down what each girl can expect to make. WeCamGirls ran an internal poll of its members, and found that while around 7 percent of them make over $5,000 a month (Vex's sort of range), half of them earn under $1,000.

So cam-models are looking for other ways to supplement their income, and find new fans. It turns out that making money from sex online goes far beyond the webcam.

UTHERVERSE

Jessica is a professional porn star who has worked in the industry for over a decade. She is also an extremely popular cam-model, and does private cam-shows every day, to a small group of high-paying subscribers.

But the majority of Jessica's clients are to be found in a virtual world called Utherverse. Utherverse is a sort of sexed-up version of Second Life. It has all the trappings of the real world: you can own your own flat (called a Zaby), bought with "Rays" that can be exchanged for American dollars. If it's your thing, you're free to settle down with a virtual family in virtual suburbia. But most people are here to meet other avatars, and for twenty-four-hour parties, nightclubs, and strip bars. Around 3,000 people join this parallel universe every day. "It's a strange mix of fantasy and reality, and that makes it the perfect place to sell sex," explains Jessica. According to Utherverse's friendly president, Anna-Lee, there are currently 25,000 avatars selling cyber-, phone-, or cam-sex in Utherverse. These avatars are easy to spot—they each have a sign that floats over their head: "working guy" or "working girl."

Anna-Lee tells me that a lot of cam-models come to Utherverse to "troll"—a term confusingly used here to refer to soliciting for business.

Along with her wife Elle, Jessica runs her own real-world multi-purpose pornography company in Massachusetts. Both are veterans of the adult entertainment industry, and both maintain a strong online and offline presence. Neither is happy with the direction that pornography is taking. "People's standards used to be higher," says Jessica. "And because the technology's gotten better, every Tom, Dick, and Harry now thinks they're a videographer or cam-model!" adds Elle. But both have adapted to this new world, and both offer a remarkable array of cam-shows, phone sex, and avatar sex, in addition to the more traditional video shoots.

You don't need to be a professional porn star to sell sex here. Anyone can become a working girl or boy and offer their services. I met Julia in one of Utherverse's carefully designed seedy brothels. As I walked in, three scantily clad women were dancing, all "working." Julia was tall, tanned, and sultry, wearing some kind of all-in-one string dress.

"Hey!" I say.

"Hey, sexy."

"So what do you do, Julia?"

"I dance, strip, and fuck, all for money," she quickly replies.

Julia is a "verified" working girl, which means she has a picture of the real her linked to her avatar. The online Julia is a tall, tanned, sultry twenty-five-year-old. The real Julia is a nurse from Kent in her mid-fifties who is happily married with five children. Julia's main business here is selling cybersex—making your avatars have sex onscreen, while typing explicit commentary into a chat box.

Julia tells me she is very good at cybersex. Some days, she has five or six customers who all want thirty minutes or even an hour of cyber-sex with her. That's three hours of work, for perhaps twenty dollars. It's not enough to provide a real-world income, but it pays for a VIP account here. She does it, she tells me, "for the Rays and the buzz."

"Doesn't that get a bit tiring, all that endless dirty talk?" I ask.

"Yeah, sometimes," says Julia.

"And does cybersex actually turn you on?" I ask.

"No, not really," she replies.

OCCUPATIONAL HAZARDS

Displaying your naked body live onscreen certainly has its downsides. Most cam-models have the occasional slow room, difficult days, and strange requests; these are just occupational hazards. "Personally, my worst nights are where no one is interacting," Blath told me, as she was getting ready for the show. The silent chat box is a cam-model's nightmare. There's no feedback mechanism. You don't even know if anyone is paying attention. "It's unnerving." Vex recalls that private shows used to be especially difficult. One fan wanted her to instruct him to take poppers.

But these are the least of a cam-model's concerns. According to Shirley, Chaturbate issues several Digital Millennium Copyright Act notices every day, because some viewers record shows and then repost them on other pornography sites, which is illegal without the site's permission. "You'll probably end up on a free porn site," Vex told me, laughing, after my short cameo. And by putting themselves onscreen, cam-models have long been a target for trolls. In August 2012, one camgirl had what appeared to be a live emotional breakdown on cam after being repeatedly trolled by users of 4chan. "God forgot I existed," she said, in tears. "Twelve years, I've been waiting for a man to love me. God doesn't care. I want to die."

And it's not only cam-models who are at risk. The growing volume of sexually explicit material that we share online or with each other has spurred a remarkable growth in what is called "revenge porn": the posting or sharing of explicit photos or videos of a person without their permission. In late 2013, Kevin Bollaert, a twenty-seven-year-old from San Diego, was arrested in relation to a revenge site he operated. It was found that he had amassed over 10,000 explicit images, all without the knowledge or permission of the subjects. Myex.com is a similar site, which was still functioning at the time of writing. Users post pictures of their ex-partners (often naked, occasionally fully clothed), accompanied by a short caption explaining why they've chosen to post: "While I was in Iraq this hoe cheated on me," wrote one. "This girl will lie to you . . . I would suggest staying away from her," wrote another. The only way to remove the pictures from myex.com was through an "independent arbitration company," who myex.com advised you to contact "if you feel you have been submitted to this site wrongfully." Removal of images would cost you $499.99. A recent civil rights report found that half of all victims of revenge porn reported that their naked photos appeared next to their full name and social network handle; 20 percent said that their emails and telephone numbers also appeared. Similar things are happening in schools, too, as cam or sexting pictures end up being shared around the class or school or friendship group. The effect is, of course, devastating. Jessica Logan from Ohio committed suicide after a nude photograph she had sent to an ex-boyfriend was shared around her school. In another American high school, a group of boys were found collecting "sexy selfies" of their fellow students, which they were using to demand ever more explicit photos from the subjects.

Vulnerable teenagers, willingly or unwillingly, can also be sucked into the world of webcams. There is a good reason why Chaturbate is so strict about verifying the age of its performers. In 2000, thirteen-year-old Justin Berry set up a webcam. Initially, he was offered $50 to take off his shirt and sit bare-chested onscreen for three minutes. He was soon being asked to pose in his underwear for a little over $100. It was the beginning of a cycle of online abuse. For over five years Berry earned thousands of dollars doing various sexual shows for hundreds of paying subscribers before the site was shut down. It was an early warning of the dangers of the webcam world.

THE CLIMAX

Vex has never met her viewers, and doesn't plan to. Her relationship with her regulars exists strictly online, a boundary she carefully polices. But part of Vex's appeal is that she is obviously real. Her shows are unashamedly home-made—a mixture of porn and, as one regular viewer describes it, a Skype chat with your girlfriend. Shirley tells me that camming is so popular because people want "the real girlfriend" experience, warts, and all. If people are going to use the internet for sexual satisfaction—and they will—camming is a more realistic and meaningful experience. Things go wrong, there are mistakes, there's chat, cats wander in and out. Vex might emphasize her ums and ahs, but she doesn't make them up. Everything is real. That's healthy. For all the social panic about the ubiquity of hard-core porn on the net, there is something quite comforting about this. The net has always been accompanied by utopian dreams of sex without limits, of fantasies without boundaries. In his famous 1990 article about the future of sex in the magazine *Mondo 2000*, Howard Rheingold argued that "the definition of Eros" would "soon be up for grabs," because everyone will be as beautiful as they want and will be able to have virtual sex with anyone, anywhere. But most people don't want fantastical sex with robots or supermodels. They want ordinary sex with real people.

Yet something about the phrase "real girlfriend experience" bothers me. I like Vex a lot. I understand why her fans keep coming back. You really do get an excellent "girlfriend experience" with her. And that's the problem. The men in her room aren't her boyfriend. Vex's boyfriend—who is a very cordial and friendly man—is currently downstairs, listening to the football on the radio.

When it's stripped down to its bare essence, camming is a transaction. This is Vex's job. Danny—a loyal Vex fan—explains the downside to me of "the girlfriend experience": "You have to keep reminding yourself that you will never meet these women in person, and *they do not want to fuck you.* Once I came to that conclusion, I found myself a much happier member of the camrooms I frequent."

Vex never says this to me directly, but I have the impression she is aware of this tension. She has come to genuinely like many of her fans—especially the regulars—some of whom she knows quite well. With one she often exchanges tips on books and new music, with another she discusses politics. She does not view her fans as cash cows. I think the reason people like Vex is not because she acts as though she genuinely cares, but because she really does. But to keep this show on the road, Vex also needs their tokens. The big, regular tippers could at any moment transfer allegiance to another model. Vex has her leaderboard of top tippers. She devises games to encourage them to transfer more tokens into her account. She treats regular spenders with particular care. The strange power dynamic might strike some people as reinforcing certain gender stereotypes: that it's okay for women to sell their body, to perform to men for money. That's not how Vex sees it—for her it's an empowering, enjoyable, and lucrative way of getting by. Yet it's not entirely clear who's really in charge here. Vex's real skill is to somehow keep this all in a workable balance. But it doesn't work for everyone. Occasionally, the uneasy model–viewer relationship explodes into direct confrontation. "You guys need to do better," wrote one well-known cam-model in an "open letter" to the viewers of MyFreeCams. "The lack of tipping and support . . . That shit needs to STOP," she fumed. "At some point, a bit of the responsibility falls on you . . . Don't throw excuses around about being unemployed, being short on cash, etc. Don't have a job? Well, stop spending your days on MFC and look for one!" One irate viewer responded: "This is a business and your job is selling your body, your personality, your services. I know it sucks sometimes. Welcome to the real world! We're not a community of philanthropists."

Tonight's show finally draws to a close. As Vex climaxes she rolls over to the side of the bed, just off-camera, and gives me a thumbs-up. It was a truly phenomenal performance by all the girls, and the fans rewarded them for it. Over 5,000 people had joined Vex's room in total, and the majority tipped, and tipped well.* The girls have earnt about £300 each for close to three hours online.

With the main event successfully accomplished, Vex shouts "Pile-on!" and the three of them jump on top of one another. "Music!" Vex laughs, from the bottom of the pile. "Let's put on some music!" The girls have just enough time left for a dance for their satisfied fans, as Duchamp and I look on.

---

* Every hour the room with the most viewers wins a ten-dollar prize. Between 8 and 9 p.m., that prize goes to Vex's room. This turns out to be one of the biggest audiences she has ever had in her room at any one time. "Three-girl shows are extremely rare," she later explained. Before the show, I asked Vex how many people she might typically have in her room. "If you're just hanging out, could be two hundred people in there, depending on how naked you are. If I'm doing a cum-show, it's usually about one thousand."

7

THE WERTHER EFFECT

"Yay! We're so glad you joined our community. You're going to love it here!" Thirteen-year-old Amelia was browsing the internet for dieting tips when this friendly message greeted her. Amelia had recently been bullied by girls at her school about her weight. She was shy, and was becoming increasingly self-conscious about her appearance. This looks nice, she thought. She clicked through.*

Three years later, Amelia was being driven to hospital by her concerned parents. She was dangerously underweight, and required urgent care. But Amelia didn't think so. "You don't understand!" she told them. "There's nothing wrong with me! This is normal. I don't *want* to recover. I'm *pro*-anorexia." At this point, she says, she was so ill that she could barely walk.

Over three years, Amelia had become a popular and committed member of the site she'd stumbled across. It was one of a

* To protect the identity of the people I mention in this chapter, I have created the composite character of Amelia. All information is derived from interviews I conducted with members of pro-ana websites, and is accurate to the best of my knowledge. I have also cloaked quotes where necessary.

large number of online "peer support" groups, websites, and forums, one dedicated to the eating disorder anorexia nervosa.

When we are feeling ill or under the weather, our first port of call is more often than not the internet. With the click or two of a mouse we can match our symptoms to a range of maladies, and quickly find individuals and communities ready and willing to offer advice and support. Today there are thousands of dedicated online peer support groups covering almost every ailment and illness imaginable, created and maintained by other sufferers, for other sufferers. Eighteen percent of U.S. internet users report having gone online specifically to seek out people with similar symptoms to their own.

Online peer support groups have been shown to help people at difficult times in their lives. Studies consistently find that speaking to people who have first-hand knowledge and experience of your own condition helps to improve self-esteem, boost confidence, and aid well-being. But the "pro-ana" site Amelia had found was part of a branch of online peer support that's less beneficial. Every day, thousands of people from all over the world visit the sprawling network of forums, blogs, and websites dedicated to various types of self-harm: anorexia, self-mutilation, suicide. Some are designed to illustrate the dangers of a particular condition, to help people to recover, or to advise them to seek help. Others are ambivalent—sites for people to speak openly and honestly about their illness. And a small minority of them are "pro."

Arguably the first "pro" self-harm site was a Usenet newsgroup called alt.suicide.holiday, or "a.s.h.," created by Californian Andrew Beals in August 1991. The first two posts on a.s.h. (now referred to as its original charter) set out its aim: "With the holidays coming up, this newsgroup will be a good resource . . . as we all know, the suicide rate raises around the holidays and this newsgroup is the place to discuss methods and reasons." a.s.h. quickly became one of the most notorious groups on the net: a place where hundreds, and then thousands, of visitors would talk about suicide, ask advice on methods to use, or even look for partners with whom to make a "pact." Today there are hundreds of similar suicide forums and sites—many of which still use the infamous a.s.h. welcome: "Sorry you're here."

By the late nineties the first pro-ana sites (and "pro-mia" sites, for bulimia) began to appear. On these sites, anorexia and bulimia were presented not as dangerous illnesses, but as lifestyle choices. The sites sought to mutually strengthen sufferers' commitment to weight loss, and provide a space to share advice and tips. According to Dr. Emma Bond, who conducted a large review into the English-speaking pro-ana community in 2012, there are between 400 and 500 main pro-ana websites and blogs on the surface web today, along with thousands of smaller blogs. "Pro-cutting" sites are also prevalent on the web, with approximately 500 dedicated sites or forums online by 2006, often linked to pro-ana sites. The number has steadily increased ever since.

The sad truth is that Amelia's case is not unique. Hundreds of people join self-harm sites each week, learning techniques and tricks, and meeting legions of like-minded individuals. A 2007 study examining the popularity of pro-ana sites found that they were visited by around half a million people. A 2011 EU study revealed that approximately one in ten eleven- to sixteen-year-olds had seen a pro-ana site. Visitors are overwhelmingly women aged between thirteen and twenty-five. a.s.h. (and a related newsgroup called alt.suicide.methods) still runs to this day, and contains thousands of threads, posts, and comments, read by an unknown number of people.

I couldn't understand how such obviously dangerous and destructive sites—sites where starving, cutting, and even killing yourself are encouraged—could be so popular, and so appealing. I went online to find out.

YAY! WELCOME!

Self-harm sites are extremely easy to find. A simple Google search reveals a number of websites, blogs, social network accounts, and image-sharing platforms all dedicated to self-harm, and all easily accessible. There is no need for special browsers or passwords.

The first pro-ana site I discover is a vast and varied multimedia experience, comprising image galleries, chat rooms, discussion forums, even an online shop for pro-ana products. Its forums include dedicated rooms for diets, relationships, physical conditions such as self-injury, and help and advice. At the time of writing, the forums alone had 86,000 members—of which 630 were online when I was visiting the site. Users create detailed profiles of themselves, including their age, location, interests. And like many other social networks, you can like and rank other people's comments, content, and profiles. Browsing through the pages, I noticed that almost every user was female, and aged between fourteen and eighteen. Alongside basic biographical information, most also featured lists of weights: a current weight, a series of weight targets, and an "ultimate" goal weight.

There is always something happening. In total, over two million comments have been made on the tens of thousands of conversation threads that have been started by users. Every two or three seconds, there is a new comment or thread being added by one of the hundreds of people online: a question for "three-to-six-a-day purgers"; a favorite diet; what do you see in the mirror?; how do you know when you're ana?; how do you hide cuts when you're at the gym? No matter what question you ask someone is there to answer: "For some reason, as soon as it reaches the evening time, it's like this switch goes off in my brain, and I want to attack myself," writes one user, which is swiftly followed by an outpouring of helpful suggestions. "Oh my gosh, thank you so much everyone," she replies.

There are also forum threads for things beyond anorexia: upsetting things that people say, songs, bad days, how to reduce skin puffiness, favorite German words, *Game of Thrones*, dating advice, dreams, pets' names, homework, dragons, and jean shorts. Rather than solely dieting tips and advice, the site provides a space for users to discuss what they like, but also, perhaps most importantly, issues only other anorexia sufferers understand. One recently created thread is titled "funny/disgusting":

> **Allbones:** This thread's for all of the yucky/funny things about your eating habits that you wouldn't discuss with anyone else . . . Let's see. The other day, I binged on spoonfuls of peanut butter and after, I was sitting in bed and burped up this terrible acid/peanut butter liquid and it was all up my throat and in my mouth . . . and I re-swallowed it with pride.
>
> **Shard:** One word. Laxatives. I was at a concert once. Right at the front and I had taken loads of lax the day

> before, I did a huge silent fart and somebody behind me heaved. Ooops.
>
> **Will-be-thin:** This one literally just made me lol. This is so embarrassing when you're in a public bathroom too with people in there because it's just like plop. plop. plop. plop. At least fifteen times ☹.

This was far from unusual. The threads are all busy, and the vast majority of comments I read are positive and encouraging.

The site is a jumping-off point. Many users add links to their own sites and platforms. The pro-ana community has always been remarkably quick to pick up and utilize the latest platforms and portals. Although it started with static websites, online journals, and Yahoo! groups in the late nineties, it quickly moved to blogs and social networking sites like Facebook and Twitter. I found hundreds of Tumblr blogs, Instagram, and Twitter accounts for pro-ana and self-harming, where users post pictures, messages, images, and videos for others to view and share.

One year after finding the pro-ana sites, Amelia joined Twitter. She noticed that a friend from the website had joined too, and was tweeting about her eating disorder. Through her, Amelia discovered a flourishing network of Twitter accounts just like her friend's. Amelia set up a new account purely for pro-ana activity. She began to tweet, and quickly became an influential member of a large network of Twitter users posting daily updates about their weight-loss efforts, and offering advice and encouragement to others, and to each other.

Amelia made several good friends online—people who sympathized with her, who always listened, and who always responded to her questions or thoughts. She began to feel like she was part of a community, and her Twitter network became increasingly important to her. "I never really talked about my ED [eating disorder] to my friends, although they knew about it, and I always hated talking to my parents," she explains. "Although they were supportive, they just didn't understand. Somehow I needed to vent my feelings to those who did. On Twitter I didn't have to hide or hold back, like I did in real life. There were times I didn't go on for a week or so because I felt so down, but then I missed speaking to other accounts. I felt like the Twitter account was a part of me. If I deleted the account or just stopped using it: then I would have just disappeared without trace."

As well as providing a space for users to discuss those aspects of a particular condition they may not wish to, or cannot, share with others, many of these sites also offer a space for users to simply talk through their problems. Following the temporary closure and reappearance of one of the largest self-injury websites in December 2013, its message board was flooded with posts from concerned users: "Rather ironically I cut more with the closure of this site! Did anyone else find this?" asked one. "I found I cut more as well haha. Really glad it's back I was checking daily aha," wrote another. "I cut LOADS more without the support I get from you wonderful people," added a third.

Gerard—a thirty-year-old American—credits a suicide forum with saving his life. Suffering from depression, he first tried to overdose at the age of eighteen, and was hospitalized. When Gerard discovered a.s.h. in 2003, he found it an enormous source of comfort. "I felt I had finally found a place I could be honest and open about my suicidal thoughts," he recalls. "To be heard and understood helped me much more than psychiatry. Putting up a front of 'being okay' for friends and family is exhausting, and makes you feel really alone. When I'm really low I check the forum a lot throughout the day. I often write long posts late at night when I feel trapped and in despair. It's always nice the next morning to read the kind and insightful replies."

Al, a moderator of a popular suicide forum, thinks that Gerard's experience is typical. His website is neither pro- nor anti-suicide. Al won't encourage anyone to take their own life: but nor will he try to talk anyone out of it. (Unlike a.s.h., he will intervene if the conversation touches on methods or making pacts, both of which are banned.) Al is a sixty-seven-year-old American, and tells me that he has been suicidal since his teens. The site, he says, has helped enormously. "I've found that just being able to talk about life with others who understand and aren't judgmental has made it much easier to not jump on the suicide train every time things go south."

Support, explains Al, comes in varied forms, and is not always what outsiders might expect. "Sometimes the best support we can give is to suggest that certain members be very cautious of their endeavors, because of what could go wrong. For others, just to write 'I do understand what you're saying!' can be enough to take the immediate pressure off. I think that since we acknowledge the right of our members to feel however they feel—and that includes feeling suicidal—and say what they like in a non-judgmental environment, it relieves, rather than encourages, suicidal tendencies." What people don't realize, Al explains, is that there is often nowhere else for these people to go.

"Running this group is not always easy," he says. "When it

becomes obvious that someone has the desire to live then it pleases me. When I have to accept that there are reasons for someone to commit suicide and I've done the best I can to provide comfort during the time they are with us, that also pleases me. I'm saddened by their death, but I can acknowledge that they're no longer in the pain that brought them here."

He must occasionally want to identify people, I suggest, to find them professional help or alert the authorities? "No. If I try to identify people then I've lost the major advantage I have. Of course, I want everyone I come into contact with to have a long, happy, loving life. But sometimes that just doesn't happen. I do feel responsible for helping every person who is active on the forum. But I'm not in the business of saving people. I'm in the business of trying to help them make the decisions that are best for them."

ENCOURAGEMENT

The stated function of almost every pro-ana site is to help visitors achieve their weight-loss goal. The busiest and most popular pages on the sites I viewed were dedicated to "thinspiration": material explicitly posted to encourage others to lose weight. "This is a great place to share your own thinspo or link to great thinspirations you see on the web. Thinspire others!" suggests one forum dedicated to thinspiration, linking to almost 30,000 photos. According to Dr. Emma Bond, thinspiration is the most common material posted on pro-ana sites and forums. Typically "thinspo" comprises photos of very slim celebrities such as Keira Knightley, Victoria Beckham, and Kate Moss, or of contributors who upload photos of themselves—albums that are created by users and then uploaded for others to view and comment on. Sometimes these are accompanied by motivational captions: "Rome wasn't built in a day, don't give up!," "Waking up thinner is worth going to bed hungry," or "That isn't your stomach rumbling—it's applauding!"

Most thinspo pictures are accompanied by praising comments from contributors, who often express their desperation to reach these impossible levels of skinniness and glamour. Under a picture of one exceptionally skinny girl I found the following:

Amazing <33

Beautiful

Love those legs. Really clear and beautiful!

Those legs are to die for.

I would give anything to look like this!

Thigh gap <33

I WANT

I'd do anything to have those legs ☹

Wish i had your body. sigh. I have a long way to go.

Man, wish i was like this

Beautiful

Could she be more perfect?

I want to have this.

So perfect

So perfect and can I ask how much you weigh??

That's what I needto look like. I can already feel my hip bones sticking out and my boyfriend can literally grab onto them, but I want to physically see those bones poking out.

Very nice!! respect

I know this girl . . . She's at my school. Everytime i see her i die a little

> inside. Yuck im so ewww :-
>
> Wish I was like that! I'm so jealous
>
> Want
>
> omg perfect!
>
> Why god oh why? You have given me fat all over and no brains but this girl gets perfection?!?!?!!? ☹ I feel huge looking at this picture.
>
> I <3 her body, so want.

Similarly, on self-injury sites—which frequently link directly to pro-ana sites—users will often post extremely graphic photographs of their self-injury with accompanying short poems, lyrics, and other images. Although many social networking sites have strict guidelines prohibiting posts and links that explicitly glamorize and promote self-harm and eating disorders, the glamorization is often indirect, built up subtly through layer upon layer of content and direct comparison.

Suicide forums tend to be slightly different, with fewer images and more discussion. But here, many users present suicide as an honorable, meaningful solution to life's problems. In a.s.h., one anonymous contributor advised another visitor who was contemplating suicide to try to enjoy the day:

> Firstly, it will be a special event regardless. Why not enjoy it as best you can. Definitely drive some place far away 40,60,100 miles or more if you can and enjoy a nice chilled-out ride on the highway or however long you enjoy driving to a nice hotel. Check in for several days just take as many long strolls as you can. You are in new energy, new people around you, that is the best thing for stimulating a new or lifesaving idea. Make a party out of it . . . If you want to chat, leave an email here or some other way to contact you. Good luck.

The great danger is that such behavior comes to be seen as normal, healthy, and appealing. Amelia would look at the thinspo pictures, comparing herself unfavorably to the glamorous photos. Being surrounded by photos and images of exceptionally skinny people led her to develop body dysmorphic disorder. Like many visitors to pro-ana sites, Amelia started to become obsessed with emaciated bodies, and with very specific signs of extreme skinniness: "thigh gaps" (a gap between the thighs when standing with the knees touching), pronounced collar bones, and jutting knee and elbow joints. According to Dr. Bond, anorexics pride themselves on achieving the remarkable mental toughness that is required to deprive yourself of food. "Many of them come to equate the feeling of hunger with happiness," Dr. Bond tells me. "I did not see anything at all wrong with a thigh gap," Amelia recalls. "It was just the thing that we all wanted. We would obsess over it." I found the following anonymous post on a pro-ana site:

> I DO NOT FUCKING NOT BELIEVE THIS. It was LITERALLY Thursday that I was looking at my thigh gap. Then WHAM. It's gone. OVERNIGHT. ACTUALLY FUCKING OVERNIGHT. It's gone. I'm totally choked. I'm so angry at myself. How did I allow this to happen? How did I let myself go so much?

Whether it's sharing experiences, uploading photographs or describing techniques and methods, the volume of data on the websites I visited is remarkable—a shared repository of knowledge on how to self-harm. That repository included a lot of very detailed advice. Amelia started to read up on weight-loss techniques, commonly known in the community as the "Ana tips"—a set of rules that, if followed, will result in drastic weight loss:

> **Rule 1:** Rules Rules Rules. This is important. You need to set rules for yourself, and if you are truly ana, you will have no problem sticking to them because you are STRONG! Rules are everything. Make your own and keep adding to them.
>
> **Rule 11:** Drink up to a shot of apple cider vinegar before eating, it's supposed to minimize fat absorption. Drinking more than a shot causes a vague nausea which helps suppress appetite.
>
> **Rule 21:** Write everything you eat and its calories. This will make you think before eating and also make you more aware of how much food and calories you are actually consuming.
>
> **Rule 27:** Press on your stomach when it grumbles. TUMS also stop stomach growling (5 calories a piece so be careful!).
>
> **Rule 34:** Never eat out of a box or jar. Always eat from a plate or bowl. This will help you in several ways: you will see how much you are really eating; you can determine in advance how much you will eat and not go back for seconds; using a small plate or bowl will make you eat even less.

On several pro-cutting forums I found advice on how to cut yourself while avoiding detection from parents or teachers. "What can I use now my family has stopped me buying razors?" asks one user. "Thin wire, staples, safety pins, a small, sharp pointed rock, a bit of glass like from a smashed light bulb, even sharp broken plastic can be used," came the helpful response. I observe the same phenomenon in action in a number of suicide forums. It is illegal in the UK to encourage or aid suicide, even if you do not know the person and do not have material involvement in the act. All that is required is the clear intention of encouraging a person to commit suicide. But providing information or discussing suicide on the internet or anywhere else is legal; providing there is no intent to encourage someone to act on it. As a result, forums like a.s.h. include a lot of information about specific ways of killing yourself. Advice on methods ranges from the very general ("I'm not looking to try any method where I could possibly scar someone like with a train . . . can you advise?") to the impossibly detailed ("I have 4 litres of High Concentrated Lime Sulfur Spray which I bought last year before it was banned. However, my car is a little bit bigger and spacier than a normal sedan and I don't want to waste my valuable supply on a failed attempt so I wanted to ask some questions . . .").

Tricks and tips are arguably the most harmful and destructive parts of these subcultures, transforming what might be vague, ill-thought-out plans into a concrete set of instructions. Every year around twenty million people attempt suicide. The majority—at least 90 percent—fail. In a study conducted by the University of Oxford Centre for Suicide Research, of 864 people who had attempted but failed to commit suicide, respondents were asked how intent they were on killing themselves. More than two thirds were moderate or low. Similarly, a 2006 survey of patients with an eating disorder found that around a third had visited a pro-eating-disorder site and 96 percent of them had learned new weight-loss methods while there. With tricks and tips you learn how to survive on under 1,000 calories a day, and many aim for 500.

ACCOUNTABILITY

Pro-ana sites are often edged with a peer pressure, as a way of encouraging each other to attain the demanding targets they have set themselves. One popular component on most pro-ana sites is "food diaries." Users will post a detailed breakdown of what they eat each day, usually accompanied by a calorie count. Many set themselves extremely punishing schedules. Publishing your plans and updating on progress, explains Amelia, is a way of keeping yourself motivated. You know others are watching and you don't want to let them down. And if you're struggling, they will encourage you.

> **Bony Queen:** This is just going to be a short, pointless post on how i did today. Really not in a good mood. I need rest, motivation and lots and lots of CIGS.
>
> Day three:
> Breakfast. Nothing.
> Lunch. One chip two small sips of milk
> Dinner. About 300 cals
>
> Day four:
> Breakfast. Nothing.
> Lunch. Two little tomatoes and sip of milk.
> Snack. 200 calorie binge on chips with lots of sour cream ☹.
> Dinner. Four fries and a McDonalds wrap. Half of a half of it. 200 calorie (guessing by amount I ate). Total. 400 But I snacked on lots of cereal and a little bread. I'm going to go with 500 maybe more.
>
> Hate this. I don't know what my total is because I can't get a grip on my snacking. I need to feel light but I just feel like I'm being pulled down by myself. I must picture tomorrow. That's the best way to start it ☹. Hope you guys are all having a better night than me. Thanks for reading.

> > **Deleted:** Don't stress too much love. I hope you are okay <3
> >
> > **xtremethin:** You can do this! just keep positive, get some fresh air, a good long sleep and see what happens tomorrow. You never know maybe tomorrow could go really well if you put your mind to it! Hope you feel better soon
>
> **Bony Queen:** Thank you both so much. You're right I do feel slightly better. I hope you make it, because you guys deserve it. Thanks again I need some encouragement it seems.:>

Wrapped up in this well-meaning supportive community, in the social interaction and feedback, are extremely destructive and unhealthy ideas and behaviors. In 2013, one popular pro-ana blogger that Amelia followed said she was embarking on a three-day fast, after eating too much over the Christmas holidays, and hoped that others would hold her to it. Within hours Amelia and dozens of others had pledged to fast with her in support.

For three days, Amelia consumed barely anything but water and ice cubes. This sort of drastic calorie reduction is extremely dangerous, and causes immediate psychological distress. In the Minnesota Starvation Experiment, which was carried out just after the Second World War, thirty-six carefully picked men, selected for their mental and physical toughness, agreed to undergo voluntary starvation. Their intake was reduced to around 1,500 calories per day—roughly half of what is considered healthy, but still far more than many anorexics consume. The men couldn't concentrate, and reported feeling socially isolated. There was a significant increase in depression, hysteria—even self-mutilation. Amelia said that the fasting was physically and mentally tough, but, at the time, seemed worth it. Not only had she lost weight, but she had visibly demonstrated her commitment to the pro-ana community, and offered support to help another ana girl.

It was a tipping point, Amelia tells me. The helpful and caring community had, imperceptibly, changed into something subtly different and far more dangerous.

THE WERTHER EFFECT

After a few weeks in the pro-ana community, explains Amelia, everything feels so *normal*. When I first visited these sites, I was shocked by the emaciated bodies, the blasé discussions about lethal cocktails or people searching for suicide pacts, the graphic photos of mutilation. This wears off very quickly. Emaciated bodies began to appear unsurprising, and ordinary. And because thinspo, tricks and tips, suicide methods and diets are put forward by a seemingly caring community of people, it is easy to forget just how deadly the advice can be. It could be said that almost any action, no matter how misguided, can quickly become acceptable—even admirable—if you believe that others are doing it too.

In 1774, the German novelist Johann Wolfgang von Goethe published his first novel, *The Sorrows of Young Werther*, in which his thoughtful young protagonist takes his life after failing in his endeavors to be with the woman he loved. The book sparked a spate of copycat suicides across Europe by young men who had found themselves in a similar predicament. This strange phenomenon became known as the "Werther Effect." The month after the August 1962 suicide of Marilyn Monroe, 197 suicides, mostly of young blonde women, appeared to have used the star's death as a model for their own. In the 1980s, a number of men in Austria committed suicide by jumping in front of trains; at the turn of the century in Hong Kong there was a spate of suicides by "charcoal burning," and in 2007–2008 in south Wales, of teenagers hanging themselves.

Sociologists call this "behavioral contagion." The Werther Effect occurs because we are social creatures. We model our behavior on others, learning from and imitating those around us. Patterns of behavior, it turns out, can often spread in much the same way as disease does. The same phenomenon has been observed with drug abuse, teenage pregnancy, self-harm, and obesity, but also with happiness and cooperation.

The Werther Effect has been found to take particular hold following cases where the victim is portrayed as romantic and heroic in some way—like Werther himself—and if they receive a lot of attention or sympathy. This is why outbreaks of the Werther Effect nearly always follow large-scale media coverage. As a result, many countries have strict guidelines about how to report suicides. During the time of the south Wales suicides, for instance, the police asked the national media to stop reporting the stories in a bid to limit the number of copycat cases.

Unlike mainstream media, there are no real guidelines or rules on reporting threats of suicide. Most suicide forums encourage users to describe how they are feeling, and why—often in an attempt to be supportive, sympathetic, and responsive, but with potentially dire results.

David Conibear was a successful computer software engineer in his late twenties, and a frequent and popular user on a.s.h. In late 1992, he posted a new comment on the site:

> Hey, fellow ASHers! . . . After a lot of research and a lot more thought, I've gone with the KCN dissolved in cold water . . . The computer is programmed to wait 36 hours and then phone 911, to prevent any of my friends discovering the body. This news message is also on a delay timer, just in case there are any closet interventionists lurking. If this DOESN'T work, I'll try to get someone to post that tidbit to a.s.h. so none of you ends up making the same mistake. Oh, a final note . . . in case the group gets any flak about this, let it be known that a.s.h. was not a promoting cause in my suicide. Had it not been for this group, my best plan to date was to get pissed drunk and dive off the roof of my apartment building (yes, I have a key). I think this is cleaner all 'round. Have a nice life!

It was the first documented online suicide note. David's body was found the next day. As the news spread on a.s.h., several users wrote short memorials, describing how saddened they were by the news, how they missed him: "Dave, if you can see this post, we're thinking about you, your spirit lives on in our minds." But a number praised him, admiring his actions: "Am I the only one who feels a perverse sense of glee when reading this letter?" wrote one user. Conibear is, somewhat shockingly, referred to as the "patron saint" of a.s.h. A micro-Werther.

The Werther Effect creates a strange and very perverse incentive, which is key to understanding how these communities can be both helpful and harmful. Because self-harm forums are generally supportive, community-minded places, the more you suffer, the more attention you get from others. Academic research has found that the motivations for self-injury and anorexia are often driven by the same underlying causes: it's a strategy to relieve feelings of anxiety, loneliness, alienation, and self-hatred. The more Amelia suffered—and expressed that suffering publicly—the more sympathy and attention she received. For someone with low self-esteem and few friends offline, this was an extremely powerful draw.

In November 2013, this incentive to perform for attention and sympathy was taken to a terrifying conclusion by a twenty-one-year-old Canadian student named Dakota. He posted a disturbing comment on 4chan's /b/ board that quickly attracted a large and captive audience: "Tonight I will be ending my own life. I've been spending the last hour making the preparations and I'm ready to go through with it. All that I request is for you guys to link me to a site where I am able to stream it."

Someone duly created a "Chateen" room, a private chat room in which Dakota could stream his webcam for the 200 or so /b/ users who had joined the thread. Dakota began streaming, and the room quickly filled with slightly sceptical /b/tards. Many suspected it was a joke. Some tried to talk him out of it; others urged him on: "You're too much of an attention whore to do it. Just fucking do it if you are gonna do it!" wrote one. "Hang yourself on school's property," suggested another.

The news spread rapidly, with thousands of other frustrated users following the developments from /b/. "Holy shit, somebody fucking stream it [the Chateen room] . . . what's wrong with you guys?!" The Chateen audience provided commentary as Dakota downed sleeping pills and gulped vodka. "Holy shit, [he] wasn't actually a faggot this time." "YES THIS MAN IS A GENIUS," commented another. Others worried: "Maybe we should work to try to save this man's life?"

By this point, Dakota had set fire to his room, and crawled under his bed. Hunched up in a ball, he managed to type: "#dead," "#lolimonfire," and finally: "IM FUCK3ED." Then the screen went dark. "I think he's passed out." No one quite knew what was happening. One user proposed a moment of silence.

Suddenly, there was a flash of light. Firefighters broke down the door and rushed in, unaware they were being filmed. They pulled the unconscious Dakota from under his bed, the large luminious yellow stripe on the trousers of one firefighter visible. "Op delivered." "He's dead. It's over." Op had delivered, but he survived. As he recovered in the hospital, his Facebook page was heavily trolled.

CONNECTED IN A LONELY WORLD

The internet hasn't created self-harming behavior. Suicide, self-harming, and eating disorders have been around for a very long time. Indeed, suicide rates in the United States have not increased dramatically since the 1960s (although they have for some sub-populations). But the net is changing how these psychological illnesses are expressed and experienced. The people who become part of these worlds are often young, extremely unwell, and in need of professional healthcare. According to the Association of Nervosa Anorexia and Associated Disorders, up to 24 million people in the United States have an eating disorder. But the reason so many join these sub-cultures is because they offer a sanctuary, when there isn't really anywhere else to go. The "Sorry you're here" welcome you receive in a.s.h. is often more than you'll receive from your local doctor. People take great comfort in being able to find and speak with others like them without judgment—and that is exactly what so many of these sites provide. It is good to have places to go to speak to other people about your suffering. Sites and forums that reduce feelings of loneliness can be extremely important when it comes to mental health problems. Academic work on the subject, although not conclusive, seems to suggest that peer support groups—if they are carefully moderated by people like Al—can help sufferers, and even nudge them towards medical help. Joe Ferns, Director of Policy, Research and Development at the Samaritans thinks that it's important for sufferers to have somewhere where they can openly talk about suicide, self-injury, and eating disorders. But he's worried by the number of untrained and often very ill people listening, offering advice and information. Everywhere there are micro-Werthers, people whose illness is glamorized and romanticized like Goethe's hero.

There's another danger too, explains Joe. Online, you can never be quite sure what advice you'll receive. Other users may be supportive, sympathetic, but there's no way of knowing who they really are. In 2008, a nineteen-year-old Canadian girl named Nadia joined a.s.h., and posted that she was feeling suicidal. A user named Cami replied. Cami explained that she also suffered from severe depression. She too had decided to end her life soon. Being a nurse, Cami could also offer some professional know-how.

Cami: I started looking for methods to let go since ive seen every method used possible at work as a emergency ward nurse i know what does and don't work so that is why i chose hanging to use ive tried it in practice to see if it hurt and how fast it worked and it was not a bad experience

Nadia: So when are you going to catch the bus?

Cami: I would like to soon you?

Nadia: I am planning to attempt this Sunday

Cami: wow ok you want to use hanging too? Or can u?

Nadia: I'm going to jump

Cami: well that is ok but most people puss out before doing that plus they don't wanna leave a terribly messy mess for others to clean up

Nadia: I want it to look like an accident. There's a bridge over the river where there's a break in the ice

Cami: ok otherwise I was gonna suggest hanging

Nadia: I considered train jumping like, at the subway, but I though this would be better

Cami: umm yeah if you wanted to do hanging, we could have done it together on line so it would not have been so scary for you

Nadia: Well if I puss out, I think we should do that

Cami: ok that sounds good im off work monday too I can die then easily or any time for that matter I w[an]t to bad

Cami: do you have a web cam?

Nadia: yes

Cami: ok well IF it comes down to hanging I can help you with it with the cam p[r]oper positioning of the rope is very important as Ive found out but we'll cross that path when/if it comes to that hun

Cami: I hope im being a help to you in some way

Nadia: yes it's a big relief to be able to talk to someone about it

Cami: I wish [w]e [b]oth could die now while we are quietly in our homes tonite :)

Nadia: since i decided that i will go this weekend, i have felt much better.

Cami: great im at peace too and if i cant die with you i will shortly after that.

Nadia: we are together in this

Cami: yes I promise

[More dialogue]

Nadia: I must say, I'm feeling a lot better now that I can talk to you

Cami: it makes me feel better too knowing I won't die alone

Nadia: you won't

Cami: Monday will be my day wish it were tonite im really at peace with it

Nadia: i wonder how it will feel to actually die

Cami: nice

In the early hours of Monday morning, Nadia told her flatmate she was going ice skating. She didn't return. Her body was discovered six weeks later. But Cami didn't go through with the pact she'd made with Nadia. In fact, Cami didn't even exist. Cami was a middle-aged man named William Melchert-Dinkel, a nurse, husband, and father from Minnesota. Police now believe he spent several years trawling the internet for suicidal individuals, and may have contacted more than one hundred people around the world in efforts to persuade them to kill themselves. Melchert-Dinkel himself believes at least five people actually did, including Nadia.

On my brief foray into the world of self-harm sites I did not discover a deviant and malicious group of people intent on causing harm to others. Although there are people like Melchert-Dinkel out there, these sub-cultures tend to be tight-knit, supportive, and caring. They are always there for you. They listen, advise, and encourage you. If you are feeling low, they are a natural and easy place to go to relieve your loneliness and suffering. That is precisely why they can be so destructive. By wrapping up negative behavior in an ordinary, positive, and romantic way—by surrounding each user with peer support—it insidiously makes an illness feel like a culture, a lifestyle choice, something to be embraced.

Eventually, Amelia was spending hours a day on pro-ana sites, posting messages about her condition, interacting with others in her community, and barely eating. She had even bought a pro-ana bracelet to wear. When her mother suggested she needed help, she refused to listen, and was terrified about losing her online social life, losing contact with the only people she thought understood her. When her parents took her to hospital, Amelia was immediately transferred to a specialist eating disorder unit. It was six months before she was discharged and finally allowed to return home.

Amelia is now fully recovered, and, for the most part, offline. I ask her what advice she might give to people who are tempted by or trapped inside pro-ana groups, as she was. "You need to get help. I know you won't want to listen. And I didn't want to listen either. But if you're actively going on these sites, you're probably already in an unhealthy state of mind. You might not think you need help, but just talk to someone anyway. There are people outside the online community who know what you're going through." She pauses. "Your pro-ana friends might understand you, but they won't help you."

CONCLUSION

ZOLTAN VS. ZERZAN

Transformative technologies have always been accompanied by optimistic and pessimistic visions of how they will change humanity and society. In Plato's *Phaedrus*, Socrates worried that the recent invention of writing would have a deleterious effect on the memories of young Greeks who, he predicted, would become "the hearers of many things and will have learned nothing." When books began to roll off Johannes Gutenberg's press, many suspected they would be "confusing and harmful," overwhelming young people with information. Although Marconi believed his radio was helping humanity win "the struggle with space and time," as his invention became popular, others feared that children's impressionable minds would be polluted by dangerous ideas and families rendered obsolete as they sat around listening to entertainment programs. We don't know if early *Homo sapiens* argued whether fire burns or warms, but you can hazard a guess that they did.

From its inception, the internet has acted as a canvas on to which we have painted positive and negative pictures of our future.

Several of the Arpanet pioneers were looking beyond data sets and communication networks to a future in which their new technology would radically transform human society for the better. Joseph Licklider, the first director of the team responsible for developing networked computing, and often referred to as the "grandfather" of the internet, predicted as much in 1961, eight years before the first network connection was made between two Arpanet nodes. "Computing," he proclaimed, "will be part of the formulation of problems . . . it will mediate and facilitate communication among human beings." It would, he believed, help us to "make better collective decisions."

Computing in the 1960s and early '70s was often endowed with a magical, mysterious power. Anarchists dreamed of a world in which humanity would be liberated from the drudge of labor, "all watched over by machines of loving grace," while countercultural writers like Marshall McLuhan were predicting a "global village" of connectedness as a result of modern media, and even a "psychic communal integration" of all humankind.

As the internet became a mainstream form of communication for millions of people there was a surge of techno-optimism. The early nineties were ablaze with utopian ideas about humanity's imminent leap forward, spurred by connectivity and access to information. Harley Hahn, an influential technology expert, predicted in 1993 that we were about to evolve "a wonderful human culture that is really our birth-right." Meanwhile the technology magazine *Mondo 2000* promised to give readers "the latest in human/technological interactive mutational forms as they happen . . . The old information élites are crumbling. The kids are at the controls. This magazine is about what to do until the millennium comes. We're talking about Total Possibilities."

Many of the net's early advocates believed that, by enabling people to communicate more freely with each other, it would help to end misunderstanding and hatred. Nicholas Negroponte—former director of the illustrious MIT Media Lab—declared in 1997 that the internet would bring about world peace, and the end of nationalism. For some, like John Perry Barlow, author of the "Declaration of the Independence of Cyberspace," this new, free world could help to create just, humane, and liberal societies—better than those "weary giants of flesh and steel."

But it was not only the optimists commenting on the possibilities presented by this strange new world. For every starry-eyed vision of future utopias there was an equally vivid dystopian nightmare. As Licklider dreamed of a harmonious world of human–computing interaction, the literary critic and philosopher Lewis Mumford worried that computers would make man "a passive, purposeless, machine-conditioned animal." In 1967, one professor warned presciently in the *Atlantic* magazine that network computing would create an "individualized computer-based federal record-keeping." As the optimism about the possibilities of the internet reached its zenith in the 1990s, so a growing number worried about the effect it was having on human behavior. In 1992, Neil Postman wrote in *The Surrender of Culture to Technology* that "we are currently surrounded by throngs of zealous Theuths, one-eyed prophets who see only what new technologies can do and are incapable of imagining what they will undo . . . They gaze on technology as a lover does on his beloved, seeing it as without blemish and entertaining no apprehension for the future."

Others were concerned that we would become "socially immature," "mentally poor," and "isolated from the outside world." Worried by the proliferation of pornography—including child pornography—and the growing amount of criminal activity taking place online, governments around the world began to pass legislation designed to monitor, control, and censor cyberspace.

This divide, between the techno-optimists and the techno-pessimists, is one that stretches back to the birth of the internet, and one that is widening as technology becomes omnipresent, faster, and more powerful. There are, today, two movements that are extreme versions of these opposing views about technology. The transhumanists embrace technology; the anarcho-primitivists reject it. Both groups have existed in some form since the early days of the internet, and both have been steadily growing in popularity in recent years, and as technology comes to play a more central role in our lives. Both exist across the dark net—from forums on the deep web to highly polished websites on the surface web, with a host of portals, blogs, and social media groups in between. But which side is right? Does connectivity bring us together, or supplant real-world relationships? Does access to information makes us more open-minded or committed to our own dogmas? Is there something about the internet, or perhaps technology itself, that shapes and constrains our choices, prodding us to behave in certain ways? And what do their prophetic visions of our technological future—one bright, one bleak—say about the dark net and how we use the internet today?

ZOLTAN

Zoltan Istvan wants to live forever. Not in the metaphorical sense—in the memory of his children, or in the words of his books—but in a very real, practical sense. And he believes that technology will soon make it possible. Zoltan is planning to upload his brain, and all its billions of unique synaptic pathways, to a computer server. "Based on current trends, I hope to upload my mind at some point around the middle of this century," he confidently informs me. Zoltan—that really is his name—is a transhumanist. He's part of a growing community of people who believe that technology can make us physically, intellectually, even morally, better. Like all transhumanists, Zoltan believes that death is a biological quirk of nature, something we do not need to accept as inevitable. Transhumanists seek the continued evolution of human life beyond its current form. They believe that we should use technology to overcome limits imposed by our biological and genetic heritage—especially mortality and physical and mental limits—thereby exceeding the constraints of the human condition, which they regard as changeable. "By thoughtfully, carefully and yet boldly applying technology to ourselves," writes Max More, a leading transhumanist philosopher, "we can become something no longer accurately described as human . . . [who would] no longer suffer from disease, ageing and inevitable death."

Transhumanism's roots are found in the ideas of science-fiction writers such as Isaac Asimov and the futurist biologist Julian Huxley, who coined the term "transhuman" in 1957. (Nick Bostrom, a well-known transhumanist, says the desire to transcend human limitations is as old as the Sumerian Epic of Gilgamesh.) Transhumanism first became prominent in California in the early nineties, the watermark period of techno-optimism. In 1993, Vernor Vinge popularized the idea of the "Singularity," the point at which artificial intelligence becomes so advanced that it begins to produce new and ever more advanced versions of itself, quickly leaving us mortals behind. Vinge hoped that transhumanists would "exploit the worldwide Internet as a combination human/machine tool . . . progress in this is proceeding the fastest and may run us into the Singularity before anything else."

By 1998, the burgeoning group came together as the World Transhumanist Association. Soon after, a number of influential transhumanists published a declaration of intent: "We foresee the feasibility of redesigning the human condition, including such parameters as the inevitability of ageing, limitations on human and artificial intellects, unchosen psychology and our confinement to planet earth." In 2008, the World Transhumanist Association was renamed Humanity+, and remains the largest formal organization of transhumanists, publishing a glossy, quarterly magazine and organizing a number of conferences and academic events. Today there are around 6,000 members from more than 100 countries—an eclectic mix of self-confessed technology geeks, scientists, libertarians, academics, and activists like Zoltan (who describes himself as a writer, activist, and campaigner all-in-one). Together they work on a dazzling array of cutting-edge technology. Everything from life extension, anti-aging, robotics, artificial intelligence (Marvin Minsky, considered one of the inventors of artificial intelligence, is a prominent transhumanist), cybernetics, space colonization, virtual reality, and cryonics. But most transhumanist technology focuses on life extension, and technological upgrades to the brain and body.

It's the possibility of a tech-powered "great leap forward" that excites transhumanists like Zoltan, who believes the possible benefits of near- and medium-term technology are too important to ignore. In addition to the personal goal of immortality, he believes synthetic biology could solve food shortages, genetic medicine may help cure diseases, bionic limbs already do transform the lives of disabled people. (Zoltan explains that, as a computer file, his carbon footprint would be greatly reduced.) They believe that connecting our brains to computer servers would dramatically increase human cognition and intelligence, which would help us solve the sort of problems we humans are likely to face in the future. For transhumanists, not to pursue every avenue to improve human capability is irrational, even a derogation of a duty to relieve suffering and improve well-being.

Dr. Anders Sandberg, a softly spoken computational neuroscientist and transhumanist, is one of the world's leading experts on "mind uploading." He is one of the few people working on how Zoltan might realize his ambition of turning his brain into a computer file. In the nineties, Anders ran the Transhumanist Society in his native Sweden, and is now a research fellow at Oxford University's Future of Humanity Institute, where he grapples with the problems of rapid human evolution.

As I meet Anders—a tall, smartly dressed man in his early forties—for lunch on Oxford's bustling high street one Saturday afternoon, I notice he is wearing a large silver medallion round his neck. It reads:

> *Call now for instructions. Push 50,000 U heparin by I.V. and do CPU while cooling with ice to 10C. Keep PH 7.5 No embalming. No autopsy.*

"It's for whoever finds me first," Anders says. "I rarely take it off when I'm out in public."

I am none the wiser.

"The critical period in cryonic freezing," he explains, "is during the first two hours or so. As soon as I'm in the nitrogen tank and my body is cooled down to 77 Kelvin, I'll be fine. The heparin is in order to help thin the blood so it doesn't clot, and so it can freeze faster."

Anders is one of 2,000 or so people around the world currently paying between £25 and £35 per month to ensure his body is preserved when he dies.* It's surprisingly little to pay for a shot at immortality. "On current trends, I estimate a 20 percent probability that I'll be woken when the science catches up," says Anders.

My first impression of Anders is of a genius but slightly madcap nineteenth-century scientist (an impression that is helped by his soft Swedish accent and precise, clipped sentences). He recently experimented with the cognitive enhancing drug modafinil, an experience that he claims was positive, and tells me he also plans to have magnets surgically inserted into his fingers so he can feel electromagnetic waves. But his main area of interest is mind uploading (what he calls "whole brain emulation"). In 2008, Anders published a 130-page instruction manual setting out exactly how the brain's content, its precise structure, pathways, and electric signals, could be transferred on to a computer chip. If it was perfectly copied, it would, thinks Anders, be indistinguishable from the real thing.

Once you've got a file, you needn't fear death—you can always be re-uploaded into a synthetic human body, or, he says, "some

* Anders will be frozen at the Alcor Life Extension Foundation in Arizona, which charges a total of $200,000 ($215,000 for UK residents) for Whole Body Cryopreservation.

kind of robot." It doesn't matter what the vessel is, according to Anders, because it would experience consciousness in exactly the same way as we do. As he describes why he thinks this is a fantastic idea, I begin to choke on my noodles, much to Anders's delight. "Ha! You see?" he laughs, as I struggle for air. "You need a back-up. Everyone needs a back-up. What a waste of human life and potential, to die choking on noodles! Ha ha!" (For a brief moment, I agree.) Ray Kurzweil—a director of engineering at Google, and probably the world's most famous transhumanist—thinks that mind uploading will be possible in 2045, as Zoltan predicted. Most mainstream scientists are less convinced by Kurzweil's estimates. (Anders is a little more conservative, which is one reason he'll be putting himself in the nitrogen tank.)

Anders explains that he now spends much of his time working through the social implications of mind uploading, rather than the technology itself. He hints at the terrifying prospect of a computer hacker being able to access your brain and change it—"we really need to start thinking about these questions," he warns, looking slightly worried. "There are as many legal, political, and social questions as there are technical ones at the moment."

Zoltan, however, is extremely excited about living for eternity as a data file. Then again, Zoltan has already packed an awful lot into his forty or so years. In his twenties, he circumnavigated the world by boat (well, almost—he made it three quarters of the way round), became a war correspondent, invented a sport called "volcano boarding," and headed a militia group in South East Asia protecting wildlife. While covering a story for the National Geographic Channel in Vietnam's demilitarized zone, Zoltan almost stepped on a landmine—his guide pushed him out of the way of the mostly buried device at the last second. "From that point I decided to dedicate my life to the transhumanist cause," Zoltan explains. He has a wife and two young children—but makes time, he says, for twelve to fourteen hours a day of transhumanist-related work. His ultimate aim, he tells me, is to live forever, or as long as possible—10,000 years or so. "If you and I were offered the chance," he tells me over Skype from his home in California, "we'd certainly try it. We'd have awe-inspiring superhuman powers."

"But what would you do?" I ask. "Ten thousand years seems like an awfully long time."

"I can only answer this based on my current brain," Zoltan patiently replies. "One day we'll have brains the size of the Empire State Building, connected to thousands of servers. The possibilities of what we could do, see, and imagine would be endless. So no, I don't think I'll get bored." He pauses. "Well, I guess I haven't been bored yet."

Transhumanists might be small in number, but they are, for the most part, extremely committed to the cause. Zoltan tells me he is planning a number of publicity stunts in the next couple of years to bring the movement to a wider audience. This includes marching with a group of robots and a large coffin to Union Square in San Francisco to protest against what he sees as a lack of government investment in life-extension science. In early 2015, he launched the Transhumanist Party—and is, at time of writing, campaigning for the 2016 presidential election. The Party advocates, among other things, to "conquer" aging—within a decade. (Zoltan knows he's not going to win, of course: but hopes to have over half a million members by the 2024 election: "that's when we'll really make our move," he says, hopefully). Many transhumanists are "biohackers"—who, like Anders, experiment with introducing new technology into their own body directly. In 2013, transhumanist Richard Lee became the first person to have implanted a headphone in his ear. In 2012, in Essen, Germany, Tim Cannon, a tranhumanist biohacker, implanted a small computer and wireless battery inside his arm. A number of American transhumanists have recently collaborated to crowdfund a "seastead," a floating community located in international waters, outside of legal jurisdiction (in 2013, they became one of the first charities to accept Bitcoin donations). Why? Possibly, thinks Zoltan, who is an ambassador to the Seasteading Institute, to escape domestic laws that prevent certain types of research such as human cloning research, which is illegal in most U.S. states, but would probably be allowed on a seastead floating in international waters. In Zoltan's recent book, *The Transhumanist Wager* (which he assures me is fiction), transhumanists launch a Third World War from their seastead "Transhumania," determined to realize their utopian plans for humanity. When I ask Zoltan quite how far he would go to pursue his philosophy, he replies: "Well . . . I would go as far as I need to go. The highest sense of morality for a transhumanist is based on how much time one has left to live. If there's very little time due to old age, disease, or war then drastic and revolutionary actions must be taken in an attempt to promote the transhumanist agenda, especially the aim of individual immortality." (Zoltan, I have come to believe, is mildly obsessed with the idea of immortality. At one point in our interview he told me that he has instructed his wife to "stick me in the freezer" if he dies unexpectedly.)

ZERZAN

In the 2014 movie *Transcendence*, Johnny Depp plays a brilliant transhumanist scientist called Dr. Caster—an Anders Sandberg type—who is building a hyperintelligent machine, in pursuit of Vernor Vinge's Singularity moment. After a TED-style talk (of course), Dr. Caster is shot by a member of a radical anti-technology terrorist group called Revolutionary Independence From Technology (RIFT). RIFT are sabotaging the work of artificial intelligence laboratories all over the world. Shooting Dr. Caster is part of the plan to disrupt what they see as the frightening march of technology.

John Zerzan thinks if the transhumanists continue on their current course, we will see the story of *Transcendence* played out on the news, rather than in cinemas. "If we keep getting closer to this so-called Singularity moment, then I think it's *highly* likely we'll be seeing anti-technology terrorists like RIFT," he tells me.

Zerzan should know. He's probably the world's most famous anarcho-primitivist, and the author of several books on why technology—from the internet all the way back to subsistence farming—is at the root of many, if not all, of today's social problems. He wants to jettison: Facebook, computers, telephones, electricity, steam-powered engines—the lot. Anarcho-primitivism is a branch of anarchist philosophy, which believes in stateless, non-hierarchical, and voluntary forms of human organization, based on simple, precivilization collective living. The most infamous neo-Luddite of modern times was the American Ted Kaczynski, better known as the Unabomber. From 1978 to 1995, Kaczynski sent sixteen bombs to targets including universities and airlines, killing three people and injuring twenty-three. In his 30,000-word essay "Industrial Society and Its Future," Kaczynski argued that his bombings were extreme but necessary to attract attention to the erosion of human freedom brought about by modern technologies requiring large-scale organization. During his trial in 1997–98, John Zerzan became a confidant to Kaczynski, offering support for his ideas, but, he is quick to make clear, condemning his actions.

Kaczynski wasn't the first. In the 1980s, the French movement Clodo (Committee for Liquidation and Subversion of Computers) firebombed a number of computer company warehouses. The Earth Liberation Front—a movement of autonomous groups dedicated to economic sabotage and guerrilla warfare to save the planet—was classified by the FBI in summer 2001 as the top domestic terrorist threat facing the country. In fact, explains Zerzan, there are already new Unabombers out there. In 2011, a New Mexico group called the Individualists Tending Toward the Wild was founded with the objective "to injure or kill scientists and researchers (by the means of whatever violent act) who ensure the Technoindustrial System continues its course." That year they detonated a bomb at a prominent nanotechnology research center in Monterrey. "We'll be seeing plenty more groups like this in the coming years if technology keeps getting faster, smarter, more intrusive," says Zerzan. "Violence against the person is not acceptable, but property destruction and genuine resistance against technological progress? Yes, this is necessary to get people's attention."

I found Zerzan via his website, the existence of which seemed a little paradoxical. "Yes, I agree," he tells me over the phone. "And I'm confronted with that dilemma every day. But ultimately, my work is about ideas. You need to use every tool at your disposal to spread those ideas, even if you dislike them." And Zerzan dislikes technology an awful lot. He recalls when he heard about the Arpanet in the 1970s, and it got him thinking about why the student protests in the 1960s hadn't achieved as much as he'd hoped. As a radical militant student at the time, he was mostly worried about civil rights and class structure. Most of Zerzan's peers thought computers were on their side.

Instead of looking forward to imagine the future, Zerzan looked back to the past, studying the early Luddite movements, and trade-union groups like the Tolpuddle Martyrs. He didn't like what he saw. "It suddenly dawned on me," he explains. "The introduction of industrial mechanization in the nineteenth century wasn't just an economic move. It was also a *disciplinary* move! It was a way to make sure that autonomous people could be controlled by capitalists." Like many techno-pessimists, Zerzan thinks technology tends to work most effectively for those who already have power, because it maintains and strengthens their grip on society's levers: more ways to watch us, control us, make us replaceable automatons just like in a nineteenth-century British factory. "The idea that technology is neutral, just a tool, is plain wrong," insists Zerzan. "That's never been the case. It embodies basic choices and values of any society."

Worse still, thinks Zerzan, we have become too dependent on technology for our everyday needs—communication, banking, shopping, etc.—and our sense of autonomy, self-reliance, and, ultimately, our freedom has been eroded as a result: "If you rely on a machine for everything, you slowly stop being a free person in any meaningful sense." For Zerzan, modern computing and the internet are the worst of all. "The internet sums up the sad cultural result of this reliance on technology we have today." Computers, he says, make you feel like you're connected with other people, but you're not. It's superficial, flighty, and distracted. By losing authentic, face-to-face communication, Zerzan thinks the internet encourages thoughtlessness, cruelty, a lack of reflection, and short attention spans. He has a point. A growing number of writers have pointed to possible long-term detrimental health effects of online stimulation, such as technostress, data asphyxiation, information fatigue syndrome, cognitive overload, and time famine.

The only answer, he says, is to leave technology behind and return to a non-civilized way of life through large-scale deindustrialization and what he calls "rewilding." If sci-fi writers like William Gibson inspire the transhumanists, the anarcho-primitivists prefer the writings of Henry David Thoreau: back to nature. I ask Zerzan how far back he's willing to go in pursuit of a natural state of existence. Should we also rid ourselves of dialysis machines? Sewage plants? Bows and arrows? He won't commit precisely on what he'd like to get rid of: he prefers to see it as a direction of travel. "We all need to start relying far less on technology. At the moment we're heading in the wrong direction, and we need to reverse it." But his ultimate vision is for us to return to what we once were, thousands of years ago: roaming groups of hunter-gatherers. "I accept, of course," says Zerzan, "this is going to be rather difficult to achieve."

Zerzan's solutions are pretty extreme. But it's not just anarcho-primitivists who are worried by a transhumanist future of boundless possibilities. Francis Fukuyama, the prominent economist who coined the expression "the end of history" to pronounce the victory of the capitalist system, has declared transhumanism the "most dangerous idea of the twenty-first century." That's probably a little unfair. One of the stated aims of Humanity+ is to think through the ethical, legal, and social implications of dramatic technological change. But the sort of rapid technological advances we're living through certainly raise several difficult questions. Scientists in Sweden are already connecting robotic limbs to the human nervous system of amputees. Panasonic will be releasing an exoskeleton suit shortly. Then there is nanotechnology, synthetic biology, the Internet of Things, algorithmic-controlled financial services, general artificial intelligence. Some of the problems this raises are existential: if Zoltan became a data file, saved on multiple servers all over the world, is he even still Zoltan? Is he still a human, deserving the same rights we accord to our species? But many of the problems are prosaic: how long should a jail sentence be if we lived to 500? Or what would be the retirement age? Who would decide who receives new technology first, and how might it be regulated?

ACORNS AND OAK TREES

On the face of it, the transhumanists and anarcho-primitivists are radically different in their views of technology. (When I asked Zerzan and Anders to debate the issue of technology via email, it collapsed after two messages.)* But although they offer radically

* Although I cannot say that Zerzan was unwilling to engage with the transhumanists. When he learned that I'd been communicating with Zoltan, he sent him the following, unsolicited message:

> I understand that you are in contact with Jamie Bartlett regarding his book project, dealing with the internet and technology more

different solutions, both Zoltan and Zerzan describe very similar problems. Both believe that we are destroying our planet, that far too much of today's suffering and hardship is preventable, and that something drastic needs to be done. They are both profoundly disappointed by what we humans have managed so far in terms of our relationship with technology, and they are both worried about the future. Ironically, one of Zoltan's biggest fears is also from the very technology that he embraces so tightly. "My only fear is that we could make machines so smart that they decide there is no use for us, and decide to wipe us out." That would rather interfere with your plan for immortality, I offer. "Yes, it would. I hope we're smart enough to control it." It is fundamentally the same fear that keeps Zerzan awake at night: what if we lose control? What if technology doesn't only shape us, but starts to control us? Zoltan would like an upgrade to ensure that we stay in charge. Zerzan would prefer to pull the plug.

It's their views about human freedom rather than technology that constitute the real dividing line between the techno-optimists

---

> generally. JB had been in touch with Anders Sandberg, who at first agreed to do a dialog with me for the end of Jamie's book. He disappeared after the "first round" of our exchange. A few years ago (2008?), producers at the *Daily Show*, American television, asked me to tape a brief debate with Ray Kurzweil and I agreed. After quite a bit of discussion of details of how and when, etc., the idea was cancelled with no explanation. It is my assumption that Kurzweil changed his mind. My question is, are you up for a public discussion or just another coward who can't back up the techno-worship you advocate? I would like a serious and widely publicised debate, of your choice of venue, etc. I'd also like a bit of funding so as to be able to come to California, which I'd think would be a good place, somewhere there(?) I await your response . . . Zerzan.

To both of their considerable credit, in November 2014, Zoltan and Zerzan had a debate at Stamford University. It can be viewed in full on YouTube.

and techno-pessimists. For the transhumanists, there is no "natural" state of man. Freedom is the ability to do anything, to be anything, to go as far as our imagination can take us. We're always changing and adapting, and embracing technology is simply the next step along the evolutionary cycle. Nothing is fixed. "Ultimately," says Anders, "I believe that humans are acorns that are unafraid of destroying themselves in order to become oak trees." We've been *Homo sapiens* for only 200,000 years or so, just a flash in the history of the earth. In a "Letter to Mother Nature," strategic futurist Max More thanks her for the bountiful endowments, but informs her that "we have decided that it is the time to amend the human constitution." Human freedom should extend to changing ourselves if we so desire. Communication via computers isn't natural or unnatural, it just is. We'll adapt to it. Zoltan accepts that there will always be people who use technology for ill—all transhumanists accept that fairly mundane point—but sees that as an unfortunate, but inevitable, part of progress. "Overall," he concludes, "the internet has brought out the best in humanity."

For the anarcho-primitivists, technology tends to distract and detract from our natural state, pushing us ever further away from what it really is to be free humans. It's freedom in a radically different sense: a freedom to be self-reliant, a freedom to be human without relying on technology. Zerzan thinks humans have already achieved the status of a mighty oak, which the transhumanists are trying to chop down and replace with a virtual simulacrum. "It's a false kind of freedom," he explains. The further from our natural state we are, the more unhappy we become. Because this freedom and power is unnatural, it's inevitable that we'll misuse it, says Zerzan. As an anarchist, he has to be optimistic about what humans can achieve when left to their own devices, but he thinks that technology has an alienating property that prevents and interrupts the natural order of things.

SHADES OF GRAY

Technology is often described as "neutral." But it could be more accurately described as power and freedom. For the transhumanists, technology provides the ability to stride across the universe, to live forever. For the anarcho-primitivists, it is a tool used to oppress and control others, to become less than human.

The dark net is a world of power and freedom: of expression, of creativity, of information, of ideas. Power and freedom endow our creative and our destructive faculties. The dark net magnifies both, making it easier to explore every desire, to act on every dark impulse, to indulge every neurosis. I came to realize that the unspoken truth about the dark net—whether it's closed groups with password barriers, or Tor Hidden Services with its drugs markets and child pornography—is that everything is close to the surface. Hidden encrypted websites and mysterious underground drugs markets sound like they exist far below the world of Google and Facebook. But cyberspace doesn't have depth. If you know where to look, everything is as accessible as everything else. In the dark net, we can simply find more, do more, and see more. And that means we have to be careful, cautious, and responsible.

The dark net fosters breathtaking creativity. The majority of the sites I visited were astonishingly adaptive and innovative. Outsiders, radicals, and pariahs are often the first to find and use technology in shrewd ways, and the rest of us have much to learn from them. At a time when most political parties are failing to capture the attention of a disenchanted electorate, a group of angry young men managed to create an international political movement in a few short months at almost no cost. Self-harm and suicide forums are filling a gap in health provision: somewhere for people with mental health problems to come together and share their experiences whenever and however they like, from the comfort of their own homes. Silk Road 2.0 was one of the most resilient, dynamic, and consumer-friendly marketplaces I've ever seen. Vex is a motivated, self-starting entrepreneur, who has found a clever way to create a successful business in the UK at a time when one in five people her age are unemployed. The Assassination Market, for all its shock value, can be seen as an ingenious and intelligent system of anonymously measuring citizen attitudes and incentivizing collective action. Their focus might be wrong or misguided, but people in the dark net use the internet in extraordinary ways. Rather than spend our energy on trying to censor, regulate, and close these sites, we would do better to learn from them, and work out how we might use the technology they have ruthlessly exploited for good.

Each individual responds differently to the power and freedom that any technology creates. It might make it easier to do bad things, but it's still a choice. Did it give succor to my darker side? Not really. It didn't make me want to self-harm, watch illegal pornography, or bully someone anonymously. I like to think that I am a well-balanced, sensible person who embarked on this experiment with my eyes open. But I did become accustomed and habituated to horrible and troubling things. It was surprising how quickly I stopped being shocked by anything. Simply put: I got used to everything. That, I realized, can be a problem. It's important to be shocked occasionally. It usefully forces us to examine our moral view. That's why people can easily get sucked into very dark and destructive places. If I had a propensity towards any of these behaviors, perhaps it would have encouraged me. For some people—for the young, the vulnerable, or the inexperienced—freedom in the dark net comes at a price. People have to be prepared for what they might encounter there.

When I first started writing this book, I had in mind something of an exposé. That I would lift the lid on the seedy underbelly of hidden internet subcultures, revealing the dangers of life online. I was prepared (maybe I even hoped?) to be indignant and outraged. I imagined this book would conclude with a series of very clear moral declarations: drug market places are unequivocally dangerous. Suicide forums are indisputably harmful. Neo-Nazis are evil. Convicted child sex offenders are beyond even a scintilla of understanding. All very black and white. All very straightforward.

That was not the case. Without exception, I left each subculture more confused and uncertain than when I'd entered. Not because everything was pleasing and uplifting—it wasn't of course—but because it was far more complicated than I'd anticipated. Where I expected moral certainty, I found ambiguity. Online drugs markets make more drugs more available to more people—yet if people are going to buy and consume drugs (and they are) sites like the Silk Road 2.0 are probably a safer way to do it. Paul the Neo-Nazi I liked very much as a person, and I saw how the net gave him a political voice—something that should be welcomed at a time of political apathy. Amir's Dark Wallet might be used by to denude the state of tax-raising powers, but it's also helping to create new secure experiments in cooperative living based on liberty and free association. These are the subtle, nuanced, and thorny moral conundrums that the dark net presents. Even after a year immersed in these questions, I'm still not entirely sure where I sit. The dark net is not black and white: it is confusing shades of gray.

In the essay "Looking Back on the Spanish Civil War," George Orwell wrote of being confronted with an enemy who was fleeing while trying to hold up his falling trousers. "I had come here to shoot at 'Fascists,'" he wrote, "but a man who is holding up his trousers isn't a 'Fascist,' he is visibly a fellow-creature, similar to yourself." Most of the chief protagonists in this book I met online first, and offline second. I always liked them more in the real world. By removing the face-to-face aspect of human interaction, the internet dehumanizes people, and our imagination often turns them into inflated monsters, more terrifying because they are in the shadows. Meeting them in person rehumanizes them again. Whether it was anarchist Bitcoin programmers, trolls, extremists, pornographers, or enthusiastic self-harmers, all were more welcoming and pleasant, more interesting and multifaceted, than I'd imagined. Ultimately, the dark net is nothing more than a mirror of society. Distorted, magnified, and mutated by the strange and unnatural conditions of life online—but still recognizably us.

ACKNOWLEDGMENTS

First and foremost, I'd like to thank all those people who have let me into their world: Paul the extreme but affable nationalist, Zack, "Old Holborn," Michael, Vex, Blath and Auryn, Amir, Pablo, Timothy May, Smári, Zoltan and Zerzan, Charlie Flowers, Tommy Robinson, Hel Gower, the anonymous EDL social media admins, @Norsefired, Jimmy Swales, Alexander Jones, Queen Lareefer, Jessica and Elle St. Claire, the strange dancer in the Utherverse brothel, Jessi, the owners of the Pink Pussy Gentleman's Club, Al the forum admin, the individuals who comprised the composite character Amelia in "The Werther Effect," Gerard and Dr. Anders Sandberg.

This book wouldn't have been possible without the dedication and talents of Tom Avery, my brilliant editor at William Heinemann, and the rest of the team there: Jason Arthur, Anna-Sophia Watts, Sophie Mitchell, Chloe Healy, Jason Smith, and Nathaniel Alcaraz-Stapleton. Thanks also to Gail Rebuck.

I relied greatly on specialists, who are quoted throughout. I'd like to reiterate my appreciation for their time. In no particular order: Fred Langford and other anonymous analysts who welcomed me to the Internet Watch Foundation, Professor Sir David Omand, Professor Sir Peter Kirstein, Emma Bond, Derek Smart, Fiyaz Mughal, Nick Lowles, Maura Conway, Mike Hearn, Khadhim Shubber, Miguel Freitas, Professor Richard Wortley, Elena Martellozzo, Tink Palmer, Nathalie Nahai, Luke Upchurch, Steve Rolles, Sam Smith, Shirley from Chaturbate, Vex's fans (especially Vince), Anna-Lee, Joe Ferns, Manko, Donald Findlater, Paul Cudenec, Celia Blay, Professor Paul Montgomery, Rachael Spellman, Beat Bullying, James Smith, Amanda van Eck from Inform at the London School of Economics, Tristan, all the members of the Oxford Transhumanist Society who welcomed me for an excellent debate in Magdalen College, especially Andrew Snyder-Beattie, Avi Roy, and Tomas Halgas.

Many people contribute to the writing of a book. The person who contributed most was undoubtedly Louis Reynolds, my fantastic researcher who worked at least as hard as I did to make this happen. Daniel Janes provided exceptional research support throughout; and Pavol Koznac, Rutger Birnie, and Joe Rowlands all produced very helpful briefing notes. And then there was a small army of friends and experts who gave their time to review various chapters and unfinished thoughts, without whom this book would look very different indeed. Above all Jonathan Birdwell, Carl Miller, Pablo Turner, Joe Miller, Catrin Nye, and Hannah Joll. They've all been indispensable. Thanks are also due to Jeremy Reffin, Professor David Weir, and the soon-to-be-Dr. Simon Wibberley, my colleagues at CASM, Eva Pascoe, Richard Boase, "General Boles," Graham Macklin, Andy Moorling,

Jake Chapman, "IamSatoshi," Ken Hollis, David Stark, Gemma Cobb, Professor Tom Boellstorff, Niels Ten Oeven, Nick Pickles, Grace, Mike Harris, Chris Waller, Sofia Patel, Phillida Cheetham, Dan Sutch, Mona Bani, Moritz Bartl, Runa Sandvik, Marley Morris, Simon Sarginson, Niki Gomez, Mevan Babaker. And my family, of course, who, in addition to reading drafts, provided all sorts of other support, including somewhere to write in peace: Daniel, Samantha, Mum (sorry for all the swearing), and Phil. Thanks also to Noreena Hertz, David Goodhart, and Catherine Mayer for early discussions about the ideas. And of course my agent, Caroline Michel, along with Rachel Mills and Jonathan Sissons at PFD who were extremely supportive and understood exactly what I was trying to do. All of my colleagues at Demos have had to endure my extended periods of absence (which they seemed to do with great pleasure). For all those who I've carelessly and thoughtlessly forgotten, please accept my apologies along with my thanks.

There are many others I cannot thank for various reasons, but to whom an enormous debt is owed. This includes the countless thousands who create and make available free software, free websites, free online archived material, often in their own time. We all owe those people a great debt. Without them I wouldn't have been able to research this book; and much else besides.

Finally to Kirsty, who first gave me the idea for this book in a very roundabout way, whose talent for writing I have tried to emulate, and who helped me throughout—albeit from a distance.

NOTES

The Dark Net necessarily relies on a large number of online sources, including forum posts, articles, and websites. A full list of working links is available at: www.windmill-books.co.uk /thedarknetlinks.

INTRODUCTION: LIBERTY OR DEATH

3 ***Tor began a life as a***: www.torproject.org/about/overview.html.en; www.fsf.org/news/2010-free-software-awards-announced.

4 ***That's why the Assassination Market***: There is an interesting parallel to be found in Ancient Greece. The word "ostracize" comes from a strange ritual that occurred every year in Athens during the fifth century BC. Each citizen would anonymously scratch the name of the person they wanted to banish from the city on to a shard of broken pottery or on a scrap of papyrus. When counted up—and assuming a certain quorum had been reached—the person who'd been named the most often would be forced to leave, "ostracized," for a decade. The fear of the vote was thought to keep everyone, especially holders

of public office, well behaved. It was democracy without justice: there was no charge, no trial, and no defense. Just a vote. As far as historians can work out, poor Hipparchos, son of Charmos, was the first person to be ostracized, for reasons now unknown.

8 ***The Pentagon hoped to create*:** The team responsible for this project was a group called the Information Processing Techniques Office (IPTO), which was part of the Advanced Projects Research Agency (ARPA) inside the Pentagon. In 1966, Robert Taylor, the IPTO boss, was funding three universities to work on something called "time sharing," which was a way to allow multiple users to access a single computer simultaneously. Each university used its own computer with its own programming language, which meant Taylor needed three teletype terminals in his office to access their work, which was infuriating and wasteful. (Taylor called it the "Terminal Problem.") He worried the problem would get worse as more of IPTO's research groups were requesting their own computers. Taylor realized the answer lay in trying to tie the computers together into a single network, allowing different computers to speak directly to each other in a common computer language. That would make it easier for researchers to share resources and results more easily. After a twenty-minute meeting with the Director of ARPA Charles Herzfeld, Taylor was given one million dollars to develop this idea. Internet seed funding—and it came from the U.S. Department of Defense. On July 3, 1969, UCLA put out a press release, "UCLA to Be First Station in Nationwide Computer Network." This story is brilliantly told in full in *When Wizards Stay Up Late.*

8 ***In July 1973, Peter Kirstein*:** P. T. Kirstein, "Early Experiences with the ARPANET and INTERNET in the UK." This new international version of the Arpanet was being called informally at the time the "Internetwork," and was shortened to the "Internet" in 1974.

9 ***Leading psychologists of the day*:** S. Turkle, *Life on the Screen*.

9 ***September 1993, the month*:** One Usenet group—www.eternal-september.org—gives the date, as of writing, as September 7247, 1993. For further information about "the September that never ends," see M. Dery, *Escape Velocity*, p. 5.

10 ***Parents panicked about children*:** textfiles.com/bbs/fever. A BBS user of the time warned others: "If you don't already own one of those evil instruments called a modem, take warning! Don't even think about buying one. Modem fever sets in very quietly; it sneaks up on you and then grabs you by the wallet, checkbook or, heaven forbid, credit cards. Eventually your whole social life relies upon only the messages you find on electronic bulletin boards; your only happiness is the programs you have downloaded. (You never try any of them, you only collect them.)"; textfiles.com/bbs/danger1.txt. As usual, the police were also way off track, seeking desperately to illustrate "warning signs" of computing obsession. The 1993 list produced by the Philadelphia police read:

> *COMPUTER ADDICTION* (WITHDRAWS FROM FRIENDS, FAMILY, ETC.) MAY LOSE INTEREST IN SOCIAL ACTIVITIES, USE OF NEW (UNUSUAL) VOCABULARY, HEAVY WITH COMPUTER TERMS, SATANIC PHRASES OR SEXUAL REFERENCE (OR SUDDEN INTEREST IN RELATED POSTERS, MUSIC, ETC.) LOOK FOR RELATED DOODLING OR WRITING. USE OF WORDS SUCH AS: HACKING, PHREAKING (OR ANY WORDS WITH "PH" REPLACING "F") LACK OF INTEREST IN SELF AND APPEARANCE OR INDICATIONS OF LACK OF SLEEP (WHICH MIGHT INDICATE LATE NIGHT MODEM-PLAY) COMPUTER AND MODEM RUNNING LATE AT NIGHT (EVEN WHILE UNATTENDED) STORING OF COMPUTER FILES ENDING IN: PCX, GIF, TIF, DL, GL (THESE ARE VIDEO OR GRAPHIC IMAGE FILES AND PARENTS SHOULD KNOW WHAT THEY ILLUSTRATE)

NAMES ON COMMUNICATIONS PROGRAMS WHICH SEEM SATANIC OR PORNOGRAPHIC, OBSESSION WITH FANTASY ADVENTURE GAMES (DUNGEONS AND DRAGONS, TRADE WARS, SEXCAPADE, ETC.).

This misunderstanding and moral panic typically accompanies most new technologies.

10 ***Whether actual or perceived, anonymity***: "The Online Disinhibition Effect," *CyberPsychology and Behaviour* 7 (3). This article was published in 2004, but Suler had set out his thesis before then, in 2001: online.liebertpub.com/doi/abs/10.1089/1094931041291295 and users.rider.edu/~suler/vita.html.

10 ***It's true that from***: J. Drew, *A Social History of Contemporary Democratic Media* (New York: Routledge; Abingdon, UK: Taylor and Taylor), p. 92; Bitnet (1980) and Fidonet (1981) soon followed, along with hundreds of smaller community networks: Cleveland Free-Net, WellingtonCitynet, Santa Monica Public Electronic Network (PEN), Berkeley Community Memory Project, Hawaii FYI, National Capitol Free-Net, and perhaps most famously of all for the nascent digital liberties movement, The WELL (1986).

10 ***Alongside purposeful and serious groups***: Bryan Pfaffenberger, "'If I Want It, It's OK': Usenet and the Outer Limits of Free Speech," *The Information Society* 12:4 (1996), p. 377.

11 ***Bell, a contributor to the list***: A. Greenberg, *This Machine Kills Secrets*, pp. 121–22.

11 ***In 1995, he set out his ideas***: There is some dispute as to whether "Assassination Politics" was first posted on the cypherpunk mailing list, or on the Usenet group alt.anarchism.

12 ***The organization that ran the market***: Bell, "Assassination Politics," part 3; web.archive.org/web/20140114101642/; cryptome.org/ap.htm. Besides, Bell added, the organization could "adopt a stated policy that

no convicted or, for that matter, even suspected killers could receive the payment of a reward . . . but it has no way to prevent such a payment from being made."

12 ***The worse the offender***: The term "Assassination Market" is never used in "Assassination Politics"; I refer to it as the most common contemporary descriptor of the system that Jim Bell proposes.

12 ***Chances are good that nobody***: "Assassination Politics," part 2.

CHAPTER 1: UNMASKING THE TROLLS

14 ***A Life Ruin***: This is a true story, which I collected and documented in full. The name has been changed, as has the date.

14 ***It was an announcement to the hundreds***: Encyclopedia 244 Dramatica—an offensive Wikipedia for trolling culture—lists camgirls as "camwhores," and describes a camwhore as "a variety of attention whore, typically a young and very stupid woman who will do anything on a webcam for attention, money, items from online wish lists, or just to be generally slutty." On 4chan and elsewhere there are several infamous camgirls. Professional camgirls are discussed in chapter 6. It's impossible to be sure how many people are ever on 4chan because the number of people viewing a page is not recorded.

15 ***The hacktivist group Anonymous?***: Users of /b/ also act responsibly, and have, in the past, worked to identify users who they believe pose a genuine threat. In 2006, one user posted on /b/: "Hello, /b/. On September 11, 2007, at 9:11 am Central time, two pipe bombs will be remote-detonated at Pflugerville High School. Promptly after the blast, I, along with two other Anonymous, will charge the building, armed with a Bushmaster AR-15, IMI Galil AR, a vintage, government-issue M1 .30 Carbine, and a Benelli M4 semi auto shotgun." Users of /b/ informed the police immediately and the poster was arrested.

19 ***One user created a fake Facebook account***: Some users were trying

to offer (pretty reasonable) advice, believing that Sarah was still on the site, "lurking." One user commented: "SARAH YES YOU ARE LURKING I'm sorry this had to happen to you, but it happens to any girl who posts nudes here. This is why girls shouldn't post nudes here. There is a board specifically for that. In the future, do not give so much information about yourself to random strangers on the internet. I know it is fun for first timers and you want to be chatty with everyone to please them, but just send a message to your friends and apologize to them because some of them will be contacted by fake profiles who will send them your nudes. Just say this to them. 'So I posted nudes somewhere on the internet. And some of you may get them from a couple of gay dudes who want to spite me. I apologize for that.' You pretty much have to make it seem like you don't give a fuck about it and have nothing to lose."

21 ***In 2007, 498 people***: www.stylist.co.uk/life/beware-of-the-troll#image-rotator-1; www.knowthenet.org.uk/knowledge-centre/trolling/trolling-study-results; www.dailymail.co.uk/news/article-2233428/Police-grapple-internet-troll-epidemic-convictions-posting-online-abuse-soar-150-cent-just-years.html#ixzz2Xtw6i21L. Section 127(1) and (2) of the Communications Act 2003 from 498 in 2007 to 1,423 in 2012; also www.theregister.co.uk/2012/11/13/keir_starmer_warns_against_millions_of_trolling_offences/.

21 ***In a poll of almost 2,000***: yougov.co.uk/news/2012/06/29/tackling-online-abuse/.

24 ***Within four years of***: Hafner, K., and M. Lyon, *When Wizards Stay Up Late*, p. 189.

25 ***Durham was attacked relentlessly and***: Ibid., pp. 216–17.

25 ***But even the first emoticon wasn't enough***: In 1982, Scott Fahlman reproposed it, as it was clearly not catching on, although nastiness

obviously was: "I propose the following character sequence for joke markers: :-) Read it sideways. Actually, it is probably more economical to mark things that are NOT jokes, given current trends. For this, use: :-(." Vertical emoticons are believed to have originated in an 1881 edition of *Puck* magazine.

26 ***Dedicated groups started to appear*:** In one 1980s user guide to flaming on BBS, the author concludes: "If American politics and advertising have taught us nothing else, they have shown that intelligence and honesty have nothing to do with being persuasive. Stated another way, personal attacks can be just as good as facts. In recognition of this universal truth, it is up to all BBS users to upgrade the quality of their 'flames' so they can take their place as a valid form of BBS communication. Remember: If George Bush can do it with Willie Horton, so can you!"

27 ***Abusers would torment the sysop*:** textfiles.com/bbs/abusebbs.txt. "The Abusing Handbook," written by "The Joker." There is no date, but the style places it in the late 1980s. It reads as if it was written by a thirteen-year-old, and the entire text is in capital letters (quoted here without corrections).

> ABUSERS DO ANYTHING TO MAKE THE BBS A WORSE THING THEN IT IS AND TO MAKE IT HARDER ON THIS SYSOP, MOSTLY THE REASONS ARE IS BECAUSE THE SYSOP IS A MAJOR ASSHOLE. THE FIRST THING IS TO DISPLAY WHAT TYPE OF THINGS ABUSING STARTS WITH LOG ON, IF YOU GET ON USING THAT TYPE OF NAME THAT I TOLD YOU ABOUT. IF THE SYSOP IS WATCHING, EITHER THEY'LL HANG UP AND LOCK YOU OUT OR BREAK IN FOR CHAT, IF THEY BREAK IN FOR A CHAT, HEARS SOME IDEAS WHICH YOU CAN SAY. 1, I'M BUSY, FUCK OFF. 2, I'M GOING TO TRASH YOU BAD! 3, LET ME OUT I HAVE ABUSING

TO DO! 4, I HAVE TO CRASH YOUR BOARD NOW, SORRY, UNOIN RULES! 5, CAN I HELP YOU! 6, CAN YOU GIVE ME SYSOP ACCESS 7, WANT TO TRY A NEW VIRUS A MADE

28 ***A 1990s Usenet troll*:** Here is a slightly extended snippet: "You are a fiend and a sniveling coward, and you have bad breath . . . You are degenerate, noxious and depraved. I feel debased just knowing that you exist. I despise everything about you, and I wish you would go away. You are jetsam who dreams of becoming flotsam. You won't make it. I beg for sweet death to come and remove me from a world which became unbearable when the bioterrorists designed you." www.guymacon.com/flame.html.

29 ***The responder would thereafter be mercilessly mocked*:** ddi.digital.net/~gandalf/trollfaq.html#item2.

29 ***In 1999, one user called "Cappy Hamper"*:** A very lengthy and valuable resource documenting early trolling is available here: captaininfinity.us/rightloop/alttrollFAQ.htm. "Dalie the Troll Betty, Joe Blow the Troll, Otis the Troll-in-Denial, and everyone from AFKMN" contributed to the document.

30 ***The Meowers began setting up*:** xahlee.info/Netiquette_dir/_/meow_wars.html.

31 ***Alt.syntax.tactical attacks were carefully planned*:** Alt.syntax's own guide to their methods was made available by users who I think may have hacked into their account, and then posted the results for others to be more aware of. Here it is: "Waves would generally break down into this kind of structure: (a) Reconnaissance (RECON): These people will go in early and usually set up camp as 'friends of the newsgroup.' They will also act as 'double-agents' to counter-flame the other waves as the invasion progresses. The key is building a bit of credibil-

ity. (b): Wave One: Wave one will usually be what starts the flame war. Those involved in this wave can go on and each have a different flame, or go on and flame in unison. They can bring in a subject of their own or flame a previous discussion. This wave calls for extreme subtlety. The quality of the flame MUST be at its highest point here. (c) Wave Two: Wave two will consist of tactics to attack the people who were sent in as recon and attempt to start totally new flame threads. The key here is that even if we attack a group of people restrained enough to resist our flame-bait, wave two will stir things up and get others to join in. (d) Wave Three: Wave three will generally change depending on the campaign, but will generally be added to push the confusion and chaos over the top. Flame the recon, flame the first wave, flame the second wave. These guys are our balls out, rude SOB's. Mop up and clean out." ddi.digital.net/~gandalf/trollfaq.html#item2.

32 ***But real life investigations into***: internettrash.com/users/adflameweb/TROLLFAQ.html.

32 ***The vicious troll, it turned out***: magstheaxe.wordpress.com/2006/08/16/memories-of-the-usenet-wars; Boyd's email to the group about Kehoe is in full here: internettrash.com/users/adflameweb/2belo.html.

33 ***Smart was also stalked by***: On April 2, 2003, Derek Smart posted the following on his website (his exasperation with the consistent trolling I think is fairly self-evident): "I have three police reports and in one such case, that kid he instigated got visited and was almost carted away. That's when they found out that he was, in fact, a juvenille. I have spoken to the San Diego police. So has my attorney. They can't do anything until he does something criminal. And they offered that we contact the FBI if we have all this evidence because it leads directly to criminal cyberstalking. We did. Nothing has become of that yet. I

tried to get a restraining order—in San Diego (I actually flew there!) but it wasn't granted because there was no implied violence or threat. I guess they're waiting for him to show up at my house and kill my family first. Especially considering that shortly after this kid mentioned spotting me in my neighborhood (he lived about 20 mins from me at the time—according to the police report), was boasting about owning a shotgun. And that was before the kid got visited and then said that he had made the whole thing up (about me calling his house, following him etc) and that Huffman had asked him to **find out where I lived.** I have almost had SEVEN years of this crap. I choose NOT to post about it nor even talk about it because some of them have been very, very painful experiences. To the extent that when this kid posted—on July 4th weekend 2000 (while I was out of town!!)—that he had seen me (he described the car I was driving, what I was wearing etc), my fiance threatened to up and leave unless we moved. So, wumpus, while somewhere obviously in your brain you have a few brain cells on the fritz and which lead you to believe that **THIS IS FUN,** let me tell you something, retard, **ITS NOT FUCKING FUN!!** I have NO idea what your motivations are and I don't give a flying fuck. You want us to turn the forum into a battleground, fine, that is EXACTLY what we'll do!"

33 ***Smart applied for restraining orders*:** The U.S. court documents relating to the charges brought by Smart against Huffman: ia700703.us.archive.org/0/items/gov.uscourts.casd.404008/gov.uscourts.casd.404008.1.0.pdf.

33 ***At the turn of the*:** SomethingAwful.com hosts a wide variety of funny and offensive content—especially blogs, videos and stories—written by editors and forum members; it also hosts several large forums. Fark.com is a satirical site with stories submitted by users of the site. Slashdot.com was more about open-source software and technol-

ogy but also had a subversive edge and was opposed to censorship. Slashdot, founded in 2000, had a vast online community, many of them Usenetters, and celebrated user-generated in-jokes and memes. SomethingAwful forum members—those who posted regularly on the site called themselves "Goons"—frequently targeted other more serious websites for raids and general mischief-making.

36 ***Out of this milieu came***: www.thestar.com/life/2007/09/22/funny_how_stupid_site_is_addictive.html.

36 ***Futaba's web address was***: jonnydigital.com/4chan-history. p. 35.

36 ***The quasi-enforced anonymity made /b/***: **Shock trolling** (v.) Shock value trolling is a common tactic practised by exposing the targeted victim to disturbing or shocking content, such as materials from shock sites, horror or pornographic images, in order to provoke a strong reaction. The Goatse image is probably the best known example (Source: Know Your Meme). **YouTube Troll** (n.) Hateful, racist, sexist, immature, misspelled, questionable comments made by internet trolls mainly consisting of an age group of 7–13, where immature coward kids go to gain confidence by writing hateful messages they'll never have the guts to say in their lifetime (Source: Urban Dictionary). **YouTube Troll II** (n.) 4chan /b/ users who randomly pick an obscure YouTube video from an obscure band, and simultaneously write serious-looking RIP messages of condolences about a band member that has supposedly just passed away. The intention is to scare fans, other band mates, family members, friends et cetera. (Disclaimer: this is actually quite funny.) (Source: Me, watching them do it). **Advice trolling** (v): Advice trolling is used to mislead people by offering dubious or false advice, especially to newbies who are less experienced and more gullible than others. Prime examples include Download More RAM, Delete System 32 and Alt*+ F4 (Source: Know Your Meme).

Bait-and-switch trolling (v.) A common tactic in online fraud and practical humor that involves falsely advertising a hyperlink as a destination of interest, when in fact, it leads to something that is irrelevant or undesirable. Examples of bait-and-switch images and videos include The Hampster Dance, Duckroll, Rickroll, Trololol, Epic Sax Guy and Nigel Thornberry Remix, as well as copypasta stories like Fresh Prince of Bel-Air, Spaghetti Stories, Tree Fiddy and Burst into Treats among others (Source: Know Your Meme). **Facebook Memorial / RIP Troll** (n.) Groups of users who look for the memorial pages on Facebook of users—especially those who have committed suicide—and then bombard the page with insults, pornography and anything else that might cause offence. www.knowyourmeme.com.

42 ***In 1990, the American lawyer*:** Remarkably, Godwin's Law was itself consciously designed to become a meme to counter the use of Nazi analogies online. archive.wired.com/wired/archive/2.10/godwin.if_pr.html.

43 ***According to other academic studies*:** A. Pease and B. Pease, *The Definitive Book of Body Language: How to Read Others' Thoughts by Their Gestures*; R. L. Birdwhistell, *Kinesics and Context: Essays on Body Motion Communication*; A. Mehrabian, *Nonverbal Communication*.

44 ***The trolls themselves claim that grief tourists*:** W. Phillips, "LOLing at Tragedy," *First Monday*: firstmonday.org/ojs/index.php/fm/article/view/3168/3115.

CHAPTER 2: THE LONE WOLF

50 ***Blood and Honor, the epicenter*:** W. De Koster and D. Houtman (2008), "Stormfront Is like a Second Home to Me." Information, Communication and Society, vol. 11, iss. 8. See also www.splcenter.org/get-informed/news/white-homicide-worldwide.

51 ***According to researchers at King's College***: J. Bergen and B. Strathern, *Who Matters Online: Measuring Influence, Evaluating Content and Countering Violent Extremism in Online Social Networks*, International Centre for the Study of Radicalisation.

51 ***In early 2007, supporters of***: O. Burkeman, "Exploding pigs and volleys of gunfire as Le Pen opens HQ in virtual world," *Guardian*, January 20, 2007; www.theguardian.com/technology/2007/jan/20/news.france (accessed December 24, 2013). The *Guardian*'s Oliver Burkeman, using an avatar, tracked the party down to Axel, another Second Life neighborhood, "where they had rebuilt their headquarters and were engaging a handful of opponents in relatively restrained debate." Wagner James Au, "Fighting the Front," January 15, 2007, *New World Notes*; nwn.blogs.com/nwn/2007/01/stronger_than_h.html (accessed December 24, 2013). The arrival of the party led to a virtual riot, in which a peaceful protest descended into a battlefield. According to Second Life authority Wagner James Au, it was a "virtual conflagration of mini-guns, cursing Frenchmen and exploding pigs." As he puts it: "And so it raged, a ponderous and dreamlike conflict of machine guns, sirens, police cars, 'rez cages' (which can trap an unsuspecting avatar), explosions, and flickering holograms of marijuana leaves and kids' TV characters, and more . . . And when the lag was not too overwhelming to stream audio, the whole fracas was accompanied by bursts of European techno . . . One enterprising insurrectionist created a pig grenade, fixed it to a flying saucer, and sent several whirling into Front National headquarters, where they'd explode in a starburst of porcine shrapnel."

51 ***The Jewish human rights organization***: web.archive.org/web/20140402122017/; hatedirectory.com/hatedir.pdf; Council of Europe, Young People Combating Hate Speech On-Line, Mapping study

on projects against hate speech online, April 2012, www.coe.int/t/dg4/youth/Source/Training/Training_courses/2012_Mapping_projects_against_Hate_Speech.pdf; Simon Wiesenthal Center, *2012 Digital Hate Report*, Simon Wiesenthal Center (accessed March 20, 2013).

51 ***The online world has become*:** C. Wolf, "The Role of the Internet Community in Combating Hate Speech," in B. Szoka and A. Marcus (eds.), *The Next Digital Decade: Essays on the Future of the Internet* (Washington, D.C.: TechFreedom). See also L. Tiven, *Hate on the Internet: A Response Guide for Educators and Families, Partners Against Hate*; www.partnersagainsthate.org/publications/hoi_defining_ problem.pdf (accessed March 20, 2013).

52 ***Large chunks of it were copied*:** A. Berwick, *2083: A European Declaration of Independence*, p. 595.

52 ***The term was popularized*:** nation.time.com/2013/02/27/the-danger-of-the-lone-wolf-terrorist/.

54 ***By June 2011, he'd farmed*:** A. Berwick, *2083: A European Declaration of Independence*, pp. 1416–18. He then goes on to say, "I just bought *Modern Warfare 2*, the game. It is probably the best military simulator out there and it's one of the hottest games this year. I played *MW1* as well but I didn't really like it as I'm generally more the fantasy RPG kind of person *Dragon Age Origins* etc., and not so much into first-person shooters. I see *MW2* more as a part of my training simulation than anything else. I've still learned to love it though and especially the multiplayer part is amazing. You can more or less completely simulate actual operations."

55 ***Some of them were supporters of*:** Breivik used the pseudonym Sigurd Jorsalfar to write on an EDL forum in 2011. He may have attended an EDL demo in 2010. www.huffingtonpost.co.uk/2011/07/26/norway-gunman-anders-brei_n_909619.html; www.newyorker.com/online

/blogs/newsdesk/2011/07/anders-breivik-and-the-english-defence-league.html; in *2083*, he claims to have had more than six hundred Facebook friends from the EDL—and he even claims to have "supplied them with processed ideological material." Tommy Robinson has repeatedly denied having any knowledge of Anders Breivik's links to the EDL.

56 ***It took these parties years***: S. Wiks-Heeg, "The Canary in the Coalmine? Explaining the Emergence of the British National Party in English Local Politics," *Parliamentary Affairs*, vol. 62, no. 3; F. McGuinness, *Membership of UK Political Parties—Commons Library Standard Note*, December 3, 2012.

56 ***Stephen Yaxley-Lennon***: "Tommy Robinson" is a pseudonym first used by a former Luton Town FC football hooligan.

56 ***Tommy and his friends decided to***: N. Copsey, *The English Defence League*, p. 8.

57 ***It attracted hundreds of people***: www.dailymail.co.uk/news/article-1187165/Nine-arrested-masked-mobs-march-Muslim-extremists-turns-violent.html.

57 ***Together with a friend***: The earliest days of the EDL remain the subject of some debate. By the time Tommy set up the EDL Facebook page, the UPL had a Facebook page ("Ban the Terrorists") with over 1,500 fans. Paul Ray, another early member, has claimed, "The original EDL was instigated by myself coming together with members of UPL (United People of Luton) and other anti-Jihad activists around the country who finally had enough of the danger posed to our local communities and the country as a whole." This is denied by Tommy, who said Ray had very little to do with the EDL in the beginning.

58 ***But the group's reputation grew***: J. Bartlett and M. Littler, *Inside the EDL*, Demos.

61 ***According to Hel Gower*:** In 2009, the EDL Facebook admins started banning people who used racist language, in response to growing media scrutiny of the group. Dozens were purged, and coalesced around another blog, mainly to complain about the touchy and politically correct admin. An old but good article is A. R. Edwards's "The Moderator as an Emerging Democratic Intermediary: The Role of the Moderator in Internet Discussions about Public Issues," *Information Polity*, 2002.

64 ***It's an online collective primarily based*:** It is now on its thirty-second incarnation, as it has been closed down so many times.

66 ***Tommy Robinson told me that almost every*:** The purpose of infiltrating a group is usually to get access to more private conversations, and then make them public. In 2012, one antifa group claimed to have found and infiltrated a hidden EDL group, The Church of the United Templars, which was being used "as a platform for grown men to post pictures of themselves dressed as Templar Knights and dream about violent attack on Muslims and 'saving England.' "

68 ***Even then, he says*:** Anders Breivik's writing also reveals how important he believed it was to ensure that nationalists make themselves difficult to identify. In *2083*, he advises: "Avoid using channels they can monitor for activities involving planning of the operation. Use aliases when corresponding while doing research. Use software which masks your IP address and other technology while researching via the internet (for example the Tor network, anonymize.net or Ipredator). Be extra careful when researching for bomb schematics (fertiliser bombs) as many terms will trigger electronic alerts. You can consider using other people's networks remotely via laptop by parking outside their apartment/house. You can also buy an anonymous laptop and browse free from your local McDonald's etc. Use software to remove

spyware, cookies etc." (*2083: A European Declaration of Independence*, p. 853).

69 ***Its self-confessed aim is to find and identify*:** People associated with Nick Lowles had managed to infiltrate the RedWatch Yahoo! group in 2004, and claim that the purpose of doxing was to subtly encourage other people to physically attack them, without actually inciting them directly.

69 ***It is infrequently updated, but retains*:** RedWatch was first published by the neo-Nazi group Combat 18 as a printed bulletin in the 1990s (probably March 1992). The website was launched in 2001. Perhaps the most significant incident, which first brought notoriety to Red-Watch, took place in April 2003, when Leeds schoolteachers Sally Kincaid and Steve Johnson had their details appear on RedWatch and soon afterwards their car was firebombed. On January 2004, the question of the legality of RedWatch was brought up by Lord Greaves, and was answered by Baroness Scotland in the House of Lords. The website was last updated on September 12, 2013, and is now only in-frequently updated; who it is that currently updates and maintains the site is unclear. www.hopenothate.org.uk/blog/insider/article/2522/redwatch-raided.

69 ***Doxers seem to know no limits*:** The Communications Act 2003, for example, makes it an offence to send an electronic communication that is grossly offensive, indecent, obscene or of menacing character. It is also an offence to use such a network to send for the purpose of causing annoyance, inconvenience or anxiety a message that the sender knows to be false. However, it is often difficult to secure pros-ecution under this piece of legislation, because of the difficulty of de-termining how serious and realistic a menace really is.

69 ***People go to great lengths*:** www.thedailybeast.com/articles/2011/01

/27/the-mujahedeen-hackers-who-clean-facebook-and-the-face book-privacy-breakthrough.html.

70 ***He was a football hooligan, and now***: In January 2014, Robinson was convicted of mortgage fraud and sentenced to eighteen months in prison. At the time of writing—June 2014—he is out on early release.

70 ***Creating our own realities is nothing new***: The American academic Eli Pariser has documented something he calls online "the filter bubble": people increasingly surround themselves with information that corroborates their own world view and reduces their exposure to conflicting information. E. Pariser, *The Filter Bubble: What the Internet Is Hiding From You*. In the UK, we already have what is called a "reality–perception gap." For example, in a 2011 survey, 62 percent of respondents thought of "asylum seekers" when asked what they associate with immigrants. In fact, asylum seekers are only 4 percent of the immigrant population. Perceptions and reality part company: and social media can make this worse. It certainly has in these groups.

CHAPTER 3: INTO GALT'S GULCH

75 ***Millions of dollars' worth of Bitcoin***: www.theguardian.com/technology/2013/apr/26/bitcoins-gain-currency-in-berlin (accessed January 9, 2014).

76 ***One day in late 1992***: R. Manne, "The Cypherpunk Revolutionary: Julian Assange" in *Making Trouble: Essays Against the New Australian Complacency*, Black Inc, p. 204. This story is also brilliantly told in A. Greenberg, *This Machine Kills Secrets: Julian Assange, the Cypherpunks, and Their Fight to Empower Whistleblowers*. I draw on Greenberg's account throughout.

77 ***They all believed that the great political issue***: S. Levy, "Crypto-rebels," www.wired.com/wired/archive/1.02/crypto.rebels.html?pg=

8&topic=, 1993 (accessed February 23, 2014); www.themonthly.com.au/issue/2011/march/1324265093/robert-manne/cypherpunk-revolutionary (accessed February 23, 2014).

77 ***At their first meeting, May set out*:** Much of this was taken from a paper May had written in 1988, titled "The Crypto-Anarchist Manifesto." At Hughes's house, the programmers were divided into two teams. One team sent around messages in anonymous envelopes trying to evade the attentions of the other group. By bouncing the envelopes around the group, they realized it was possible to send a message without anyone working out who it originated from.

77 ***But computer systems could*:** Quoted in S. Levy, Crypto: *How the Code Rebels Beat the Government—Saving Privacy in a Digital Age*, p. 208. In 1991, Gilmore said, "I want to guarantee—with physics and mathematics, not laws—things like real privacy of personal communications . . . real privacy of personal records . . . real freedom of trade . . . real financial privacy . . . real control of identification." One early post from the mailing list gives a very good flavor of the mood: "The people in this room hope for a world where an individual's informational footprints—everything from an opinion on abortion to the medical record of an actual abortion—can be traced only if the individual involved chooses to reveal them; a world where coherent messages shoot around the globe by network and microwave, but intruders and feds trying to pluck them out of the vapor find only gibberish; a world where the tools of prying are transformed into the instruments of privacy." In his comment about democracy not providing lasting freedom, May was in fact quoting fellow cypherpunk Mike Ingle: koeln.ccc.de/archiv/cyphernomicon/chapter16/16.5.html.

78 ***The list was hosted by the server*:** Levy, *Crypto*. Toad.com is one of the first one hundred .com domain names.

78 ***Tim May proposed, among other***: He may even have been the first to write about the branch of steganography called "Least Significant Bit" in which messages are hidden in parts of audio or video files, in sci.crypt mailing list, unfortunately now lost.

78 ***When Hughes put forward a program***: From Tim May, *Cyphernomicon*: "The Cypherpunk and Julf/Kleinpaste-style remailers were both written very quickly, in just days; Karl Kleinpaste wrote the code that eventually turned into Julf's remailer (added to since, of course) in a similarly short time."

79 ***It was Hughes who coined***: www.activism.net/cypherpunk/manifesto.html (accessed February 23, 2014).

79 ***Public Key encryption transformed***: Tim May offers an explanation in *Cyphernomicon*: "I did find a simple calculation, with 'toy numbers,' from Matthew Ghio: 'You pick two prime numbers; for example 5 and 7. Multiply them together, equals 35. Now you calculate the product of one less than each number, plus one. (5-1) (7-1)+1=21 [sic]. There is a mathematical relationship that says that x = x^21 mod 35 for any x from 0 to 34. Now you factor 21, yields 3 and 7. You pick one of those numbers to be your private key and the other one is your public key. So you have: Public key: 3 Private key: 7 Someone encrypts a message for you by taking plaintext message m to make cyphertext message c: c=m^3 mod 35. You decrypt c and find m using your private key: m=c^7cmod 35. If the numbers are several hundred digits long (as in PGP), it is nearly impossible to guess the secret key.' " (The calculation is actually incorrect: when I asked him, May explained that Cyphernomicon was only a first draft, and that he'd never gotten around to checking it as carefully as he would have liked.) David Kahn, a historian of cryptography, called this the most important cryptographic development since the Renaissance. Also K. Schmeh,

Cryptography and Public Key Infrastructure on the Internet.

80 ***"Before PGP, there was no way"***: Interview with Zimmermann, *InfoWorld* magazine, October 9, 2000, p. 64.

81 ***In the end, they decided against***: In fact, three GCHQ mathematicians had already invented public key encryption a few years before Hellman and Diffie, but GCHQ chose to keep it secret. When he became GCHQ Director in 1996, Omand decided to publicly release their original proofs.

82 ***In 1994, May published***: *Cyphernomicon* began: "Greetings Cypherpunks, The FAQ I've been working on for many months is now available by anonymous ftp, details below. Because there is no 'official' Cypherpunks group, there shouldn't be an 'official' Cypherpunks FAQ, as I see it. Thus, others can write their own FAQs as they see fit. Cypherpunks write FAQs? I've decided to give my FAQ a name, to prevent confusion. 'THE CYPHERNOMICON' is what I call it. (If the reference is obscure, I can explain.)"

82 ***The cypherpunks were advised to read***: www.mail-archive.com/cypherpunks@cpunks.org/msg00616.html; Levy, Crypto, p. 207. Hughes, in his own version of *Cyphernomicon*, wrote that "with the right application of cryptography, you can again move out to the frontier—permanently."

83 ***Dyson replied, "For the record"***: www.themonthly.com.au/issue/2011/march/1324265093/robert-manne/cypherpunk-revolutionary. Assange's original posts are still preserved on the Cypherpunk list archive, which is available here: cypherpunks.venona.com/.

83 ***"I count him as one of us"***: For more on the importance to Assange of the Cypherpunk mailing list, see A. Greenberg, *This Machine Kills Secrets*, p. 127, Manne, R., pp. 207–13. (Assange went as far as to publish a book in 2012 called *Cypherpunks*.)

84 ***The experience, he later wrote***: www.themonthly.com.au/issue/2011/march/1324265093/robert-manne/cypherpunk-revolutionary (accessed February 23, 2014).

84 ***His inspiration came from another cypherpunk***: It now hosts 70,000 documents, including the names of CIA and MI6 agents, suppressed photos of soldiers killed in Iraq, and maps of government facilities.

84 ***We intend to place a new star***: A. Greenberg, *This Machine Kills Secrets*, p. 131.

84 ***It was finally discontinued in 2001***: www.securityfocus.com/news/294.

86 ***It is currently run and managed***: cooperativa.cat/en/whats-cic/background/; www.diagonalperiodico.net/blogs/diagonal-english/from-critique-to-construction-the-integrated-cooperative-in-catalonia.html.

86 ***CIC's vision is to find new ways***: G. D'Alisa, F. Demaria, and C. Cattaneo, "Civil and Uncivil Actors for a Degrowth Society," *Journal of Civil Society*; www.tandfonline.com/doi/pdf/10.1080/17448689.2013.788935.

87 ***Throw in the communal cooking***: "Degrowth in Action," from *Opposition to Alternatives Building: How the Cooperative Integral Catalana Enacts a Degrowth Vision*. It is the 2012 master's thesis of Sheryle Carlson, of the Human Ecology Divison of Lund University.

88 ***In 2009, Duran began promoting***: enricduran.cat/en/i-have-robbed-492000-euros-whom-most-rob-us-order-denounce-them-and-build-some-alternatives-society-0/.

90 ***Although Amir's technical know-how***: bitcointalk.org/index.php?topic=169398.0 (some also blame him for the 2011 Bitcoinia controversy, when a Bitcoin trading exchange partly run by Amir was hacked and £145,000 in Bitcoins were stolen).

90 ***"We believe this is not in"***: www.forbes.com/sites/andygreenberg/2013/10/31/darkwallet-aims-to-be-the-anarchists-bitcoin-app-of-choice/.

91 ***Both he and Cody Wilson***: www.wired.co.uk/news/archive/2014-04/30/dark-wallet/.

91 ***Although he never attended a meeting***: S. Levy, *Crypto*, pp. 216–17. D. Akst, "In Cyberspace, Nobody Can Hear You Write a Check: Cash? History, The Evolution of Money is Moving Way Faster Than the ATM Line. Guard Your Passwords," *Los Angeles Times*, February 4, 1996.

93 ***He even added an out-of-place***: en.bitcoin.it/wiki/Genesis_block (accessed January 9, 2014).

94 ***In his early posts, Satoshi wrote***: www.mail-archive.com/cryptography@metzdowd.com/msg10001.html (accessed January 9, 2014). Many of the most influential people in the development of Bitcoin in these mailing list days—Wei Dai, Nick Szabo, Adam Back and, of course, Hal Finney—were all veterans of the Cypherpunk mailing list.

94 ***Satoshi typed in his last post***: Until 2014, when a *Newsweek* journalist claimed to have found him living quietly and humbly in California. But the man *Newsweek* identified claims Bitcoin has nothing to do with him.

95 ***The Dark Wallet will include a number of new features***: Although some users are sceptical about how well Dark Wallet will work, because it is a full reimplementation of the Bitcoin protocol, which is highly ambitious, even for Amir.

95 ***One of the key innovations***: wiki.unsystem.net/index.php/DarkWallet/Multisig.

95 ***Amir anticipates that a lot of people***: In May 2014, an alpha version of the Dark Wallet was made publicly available: Amir encouraged users to test the software as he continued to work on it.

98 ***One is a social media platform***: Technically speaking, Twister doesn't store the posts themselves into the blockchain, but rather just the username records.

101 ***So Smári and two colleagues***: www.indiegogo.com/projects/Mailpile-taking-email-back.

101 ***More and more people are starting***: www.dailydot.com/news/pgp-encryption-snowden-prism-nsa/.

101 ***In 2013, documents released by Edward Snowden***: James Ball, Julian Borger, and Glenn Greenwald, "Revealed: how United States and UK spy agencies defeat internet privacy and security," *Guardian*, September 6, 2013; www.theguardian.com/world/2013/sep/05/nsa-gchq-encryption-codes-security (accessed November 20, 2013). Ellen Nakashima, "NSA has made strides in thwarting encryption used to protect Internet communication," *Washington Post*, September 5, 2013; articles.washingtonpost.com/2013-09-05/world/41798759_1_encryption-nsa-internet (accessed November 20, 2013).

102 ***By the next morning***: www.huffingtonpost.co.uk/eva-blumdumontet/cryptoparty-london-encryption-_b_1953705.html (accessed February 23, 2014).

102 ***There is even a free crypto-party handbook***: It is available here: github.com/cryptoparty/handbook (accessed February 23, 2014).

102 ***I've documented at least***: besva.de/mirror-cryptoparty.org/ (accessed February 23, 2014). Obviously there is no central authority for the crypto-party movement. Its reach is likely to be great because each participant is expected to take what they've learned and share it with others, possibly having their own private crypto-party. Encryption takes two, after all.

103 ***Surveys consistently show that we value privacy***: J. Bartlett, *Data Dialogue*.

107 ***Crypto-currencies can "help"***: enricduran.cat/en/statements172013/.

108 ***A place where citizens can exist***: In a 1995 email to transhumanists who were planning to build a floating seastead community to live outside national laws, May urged them to think about computer networks instead, which he considers more hospitable and secure than any physical location—even one in the ocean.

CHAPTER 4: THREE CLICKS

111 ***I contacted the police***: Given the sensitive nature of the subject, it is worth spending a moment on definitions. The seminal psychiatric document, *The Diagnostic and Statistical Manual* (fourth edition, text revised—DSM-IV-TR) of the American Psychiatric Association, has a specific definition of pedophilia: the individual must have experienced intense and recurring sexual fantasies involving children over a minimum period of six months, or various behaviors or urges involving sexual activities with a prepubescent child or children. This individual will also experience significant impairment or distress in social, occupational or other functions due to the presence of these fantasies. Finally, a pedophile must be at least sixteen and at least five years older than their victims. (The new *DSM-V*—released 2013—kept its definition exactly the same, except for a routine name change from "pedophilia" to "pedophilic disorder.") Child sex offending is different as it can cover people who are convicted for a range of criminal offences, including viewing illegal material. (The DSM's definition of pedophilia has not been without controversy: critics have argued that it pays insufficient attention to factors such as pedophiles' inability to control themselves, and also fails to distinguish "hebephiles," i.e., offenders who are exclusively attracted to pubescent children.) For the purposes of this chapter, I use the terms child pornography and

indecent images of children interchangeably. Specialists in the field prefer the term "child sexual abuse images," because all illegal images of young people can be fairly described as sexual abuse. However, although many images are clearly "abuse" in the sense most people would understand it, the term can also be slightly misleading for a non-specialist, as it gives the impression that there is always direct physical abuse involved, which is not the case. The definition cited in the text is from the Office of the United Nations High Commissioner for Human Rights (2002).

111 ***The internet has radically changed*:** There are three different ways in which people engage in sexual offences of children online, which are often incorrectly conflated. There are those who view, collect and distribute child pornography. There are others who engage in "virtual" abuse, in which an adult has some kind of sexual relationship with a child online, possibly involving webcams or exchanging images, but never meet them in person. Finally, there are those who use the internet to find and groom children, with the intention of meeting them. The relationship between these different types of abuse is murky: some offenders commit all three offences, others only the first or second.

111 ***Most countries*:** In the UK, this is called the Sentencing Advisory Panel scale. Level 1: Nudity or erotic posing with no sexual activity. Level 2: Sexual activity between children, or solo masturbation by a child. Level 3: Non-penetrative sexual activity between adult(s) and child(ren). Level 4: Penetrative sexual activity between child(ren) and adult(s). Level 5: Sadism or bestiality. In April 2014, this was changed to three levels. Category A (Level 4 & 5); Category B (Level 3) and Category C (defined as those not falling into Categories A or B). See Sentencing Council Sexual Offences Definitive Guidelines (2014).

111 ***During the sexual liberation movement*:** T. Tate, *Child Pornography: An Investigation*, pp. 33–34.

111 ***By the late seventies*:** T. Tate, p. 33; P. Jenkins, *Beyond Tolerance: Child Pornography on the Internet*, p. 32. There was also surge in pro-pedophile pressure groups, which publicly called for the legalization of sexual relationships between adults and minors. In the UK, the Paedophile Information Exchange (PIE) was founded in 1974 and even became an affiliate of the National Council for Civil Liberties. And most famously the North American Man/Boy Love Association (NAMBLA), established in 1978, still campaign today. I. O'Donnell and C. Milner, *Child Pornography: Crime, Computers and Society*, p. 11; S. Ost, *Child Pornography and Sexual Grooming: Legal and Societal Responses.*

112 ***In the UK, many pedophiles*:** This information is based on an interview with a specialist who works with recovering sex offenders, and who has asked to remain nameless.

112 ***Because it was so hard to come by*:** Quoted in R. Wortley and S. Smallbone, *Internet Child Pornography: Causes, Investigation and Prevention.*

112 ***In 1993, Operation Long Arm targeted*:** articles.baltimoresun.com /1993-09-01/news/1993244018_1_child-pornography-distribution-of-child-computer. One FBI officer involved in these raids said, "Of all the techniques used by child pornographers, none has been more successful than the worldwide use of BBSs."

112 ***Anonymous Usenet groups*:** P. Jenkins, *Beyond Tolerance: Child Pornography on the Internet*, p. 54.

113 ***Prospective members had to be put forward*:** Wortley and Smallbone, Internet Child Pornography, p. 66.

113 ***Seven UK men were convicted*:** K. Sheldon and D. Howitt, *Sex Of-*

fenders and the Internet, p. 28; www.theguardian.com/uk/2001/feb/11/tracymcveigh.martinbright and news.bbc.co.uk/1/hi/uk/1166643.stm.

113 ***The infamous Lolita City***: This is taken from an anonymous letter published online and available in a Tor Hidden Service, from someone who claimed to be a member of an international child pornography ring ("Mr X"), a German living abroad, "where there are no laws relating to surfing, viewing, downloading and saving any type of files." He claims to have worked in the field of children's models "and knows hundreds of pedophiles."

113 ***By October 2007, Interpol's***: Cited by I. A. Elliott, A. R. Beech, R. Mandeville-Norden, and E. Hayes, "Psychological profiles of internet sexual offenders: Comparisons with contact sexual offenders," *Sexual Abuse*, 21, pp. 76–92.

113 ***By 2010, the UK police database***: www.official-documents.gov.uk/document/cm77/7785/7785.pdf; www.bbc.co.uk/news/uk-21507006; D. Finkelhor and I. A. Lewis, "An Epidemiologic Approach to the Study of Child Molesters," in R. A. Quinsey and V. L. Quinsey (eds.), *Human Sexual Aggression: Current Perspectives* (Annals of the New York Academy of Sciences); G. Kirwan and A. Power, *The Psychology of Cyber Crime: Concepts and Principles*, p. 115; www.theguardian.com/lifeandstyle/2013/oct/05/sold-mum-dad-images-child-abuse.

113 ***In 2011, law enforcement authorities***: www.justice.gov/psc/docs/natstrategyreport.pdf and www.ussc.gov/Legislative_and_Public_Affairs/Public_Hearings_and_Meetings/20120215-16/Testimony_15_Collins.pdf.

113 ***Twenty-five years on from the NSPCC's estimate***: Wortley and Smallbone, Internet Child Pornography.

114 ***CEOP believes that there are*:** CEOP (2013) Threat Assessment of Child Exploitation and Abuse.

114 ***According to the same source*:** This was a hacker named Intangir—who reputedly also runs another infamous Tor Hidden Service known as "Doxbin," where the personal details of many anonymous users are listed.

114 ***One academic has recorded nine*:** T. Krone, "A Typology of Online Child Pornography Offending," *Trends & Issues in Crime and Criminal Justice*, no. 279. The profile of sex offenders has been the subject of a lot of academic work. See D. Grubin, "Sex Offending Against Children: Understanding the Risk," *Police Research Series* 99, p. 14; E. Quayle, M. Vaughan, and M. Taylor, "Sex offenders, internet child abuse images and emotional avoidance: The importance of values," *Aggression and Violent Behaviour*, 11, pp. 1–11; K. C. Siegfried, R. W. Lovely, and M. K. Rogers, "Self-reported Online Child Pornography Behaviour: A Psychological Analysis," *International Journal of Cyber Criminology* 2, pp. 286–97; D. L. Riegel, "Effects on Boy-attracted Pedosexual Males of Viewing Boy Erotica," *Archives of Sexual Behavior* 33, pp. 321–23; J. Wolak, D. Finkelhor, and K. J. Mitchell, *Child-Pornography Possessors Arrested in Internet-Related Crimes: Findings from the National Juvenile Online Victimization Study* (National Center for Missing and Exploited Children); L. Webb, J. Craissati, and S. Keen, "Characteristics of Internet Child Pornography Offenders: A Comparison with Child Molesters," *Sexual Abuse* 19, pp. 449–65; I. A. Elliott, A. R. Beech, R. Mandeville-Norden, and E. Hayes, "Psychological Profiles of Internet Sexual Offenders: Comparisons with contact sexual offenders," *Sexual Abuse*, 21, pp. 76–92.

116 ***"Legal teen" content has always been*:** O. Ogas and S. Gaddam, "A Billion Wicked Thoughts," pp. 21–28.

116 ***According to the Internet Adult Films Database***: gawker.com/5984986/what-we-can-learn-from-10000-porn-stars.

116 ***The three most commonly requested***: This does not mean necessarily that thirteen is the most popular age overall: people are more likely to specify a very exact age if they have an interest in illegal pornography. If your preference is for adult pornography, you probably are not interested in specific age categories.

118 ***According to research conducted by the charity***: www.lucyfaithfull.org.uk/files/internet_offending_research_briefing.pdf.

119 ***Another academic study found***: B. Paul and D. Linz, "The Effects of Exposure to Virtual Child Pornography on Viewer Cognitions and Attitudes Toward Deviant Sexual Behavior," *Communication Research* 2008, vol. 35, no. 1, pp. 3–38.

119 ***The authors suggest that this is***: O. Ogas and S. Gaddam, "A Billion Wicked Thoughts," pp. 176–77.

120 ***She explains to me that offenders***: E. Martellozzo, "Understanding the Perpetrators' Online Behaviour," in J. Davidson and P. Gottschalk, *Internet Child Abuse: Current Research and Policy*, p. 116. See also E. Martellozzo, *Grooming, Policing and Child Protection in a Multi-Media World*; G. G. Abel, J. Becker, et al., "Complications, Consent and Cognitions in Sex Between Adults and Children," *International Journal of Law and Psychiatry* 7, pp. 89–103; S. M. Hudson and T. Ward, "Intimacy, Loneliness and Attachment Style in Sex Offenders," *Journal of Interpersonal Violence* 12(3), 1997, pp. 119–213; E. Martellozzo, pp. 118–19.

120 ***One important aspect of John Suler's***: J. Suler, "The Online Disinhibition Effect," *CyberPsychology and Behaviour*.

120 ***Users of these "legal only" forums***: E. Martellozzo, "Children as Victims of the Internet: Exploring Online Child Sexual Exploitation," forthcoming.

122 ***Most remarkably NAMBLA considers***: I tried to contact NAMBLA via email, but unsurprisingly they replied saying that they would not respond to my questions.

122 ***Notwithstanding the gravity of possessing***: G. Kirwan and A. Power, *The Psychology of Cyber Crime: Concepts and Principles*, p. 123.

122 ***Many internet sex offenders***: Sheldon and Howitt, p. 232.

123 ***In fact, in the United States, data aggregated***: D. Finkelhor and L. Jones, "Has Sexual Abuse and Physical Abuse Declined Since the 1990s?" www.unh.edu/ccrc/pdf/CV267_Have%20SA%20%20PA%20Decline_FACT%20SHEET_11-7-12.pdf; www.nspcc.org.uk/Inform/research/findings/howsafe/how-safe-2013-report_wdf95435.pdf. However, it is notoriously difficult to draw hard and fast conclusions from data sets like this. Tink Palmer suggested that changes in the way that child exploitation is recorded may account for the data; www.nspcc.org.uk/Inform/research/statistics/comparing-stats_wda89403.html.

123 ***Despite fears about online predators***: D. Boyd, *It's Complicated: The Social Lives of Networked Teens*; www.safekidsbc.ca/statistics.htm; www.nspcc.org.uk/Inform/resourcesforprofessionals/sexualabuse/statistics_wda87833.html.

123 ***According to Peter Davies***: www.telegraph.co.uk/technology/facebook/10380631/Facebook-is-a-major-location-for-online-child-sexual-grooming-head-of-child-protection-agency-says.html; ceop.police.uk/Documents/strategic_overview_2008-09.pdf.

123 ***A CEOP and University of Birmingham study***: www.bbc.co.uk/news/uk-21314585.

123 ***Eighty of them became virtual friends***: E. Martellozzo, "Understanding the Perpetrators' Online Behaviour," pp. 109–12.

124 ***They would slowly try to build***: E. Martellozzo, "Children as Victims of the Internet."

124 ***"They spend hours monitoring":*** L. A. Malesky, "Predatory Online Behaviour: Modus Operandi of Convicted Sex Offenders in Identifying Potential Victims and Contacting Minors Over the Internet," *Journal of Child Sexual Abuse* 16, pp. 23–32; J. Wolak, K. Mitchell, and D. Finkelhor, "Online Victimization of Youth: Five Years Later," *National Center for Missing and Exploited Children Bulletin*; www.unh.edu/ccrc/pdf/CV138.pdf.

125 ***Many online groomers are extremely cautious*:** (Incidentally, as young people share an increasing amount about themselves online, creating a believable and authentic fake profile for police sting operations becomes more complicated. Your fake person now needs a fake network too, full of friends, interests and history.)

125 ***One had even posted naked photos*:** E. Martellozzo, "Understanding the Perpetrators' Online Behaviour," p. 107.

128 ***In 2006, the IWF registered*:** The 2006 report is available online: www.enough.org/objects/20070412_iwf_annual_report_2006_web.pdf.

128 ***In 2013, this was down to*:** www.iwf.org.uk/resources/trends.

128 ***Today an image could be created*:** Quoted in Wortley and Smallbone, *Internet Child Pornography*.

128 ***The overwhelming majority of the material*:** This is partly a result of how the IWF works. Obviously accidental finds are more likely on the surface web. The 600,000 people looking for child pornography in the deep web are unlikely to phone up the IWF.

130 ***After Freedom Hosting was taken*:** motherboard.vice.com/blog/the-fbi-says-it-busted-the-biggest-child-porn-ring-on-the-deep-web-1.

131 ***In 1999, the FBI seized*:** A married couple, Thomas and Janice Reedy, were convicted of trafficking child pornography via Landslide in 2001; www.pcpro.co.uk/features/74690/operation-ore-exposed.

133 ***Pornography of all types is now*:** shareweb.kent.gov.uk/Documents

/health-and-wellbeing/teenpregnancy/Sexualisation_young_people.pdf; p. 45. www.childrenscommissioner.gov.uk/content/publications/content_667.

133 ***Although data is highly variable***: J. Ringrose, R. Gill, S. Livingstone, and L. Harvey, "A Qualitative Study of Children, Young People and Sexting," NSPCC; www.nspcc.org.uk/Inform/resourcesforprofessionals/sexualabuse/sexting-research-report_wdf89269.pdf.

133 ***According to the IWF***: "Threat Assessment of Child Sexual Exploitation and Abuse" (PDF): CEOP. Other statistics put it slightly lower, but still at around 20 percent; www.pewinternet.org/Reports/2013/Teens-Social-Media-And-Privacy.aspx.

CHAPTER 5: ON THE ROAD

135 ***Approximately 50 percent of all global***: Nielsen Global Digital Shopping Report, August 2012; fi.nielsen.com/site/documents/Nielsen-GlobalDigitalShoppingReportAugust2012.pdf (accessed April 19).

135 ***According to a 2014***: See *The Global Drugs Survey 2014*. This statistic is supported by the *Guardian*'s 2014 British Drugs Survey. Based on a survey of around 1,000 British adults conducted by the polling company Opinium Research, only 2 percent of British adults who have ever taken drugs have bought them via the Internet. However, that figures jumps to 16 percent among current users.

136 ***Using the Arpanet account***: J. Markoff, *What the Dormouse Said: How the Sixties Counterculture Shaped the Personal Computer Industry*, p. 75.

136 ***According to an FBI indictment***: www.wired.com/images_blogs/threatlevel/2012/04/WILLEMSIndictment-FILED.045.pdf; M. Powers, *Drugs 2.0*, chapter 9, "Your Crack's in the Post."

136 ***With a sophisticated traffic encryption system***: In fact, Silk Road was

only one site among many. In June 2011, Black Market Reloaded was founded. While Silk Road had a few restrictions on sale, Black Market Reloaded would sell anything. Others followed: the Russian Anonymous Market Place (2012), Sheep Market (February 2013), Atlantis Online (March 2013—again announced on bitcointalk). Academics from the University of Luxembourg have recently conducted a clever analysis of Tor Hidden Services. They located around 40,000 sites, the majority of them in English. Adult content—and a proportion of that will be child pornography—accounts for 17 percent of the sites; drugs 15 percent; counterfeit goods 8 percent; and hacking information 3 percent. However, 9 percent of all sites they found are about politics, 7 percent cover hardware-/software-related subjects, and 2 percent are about art. There are also sites for games, science and sports. Given the nature of Tor Hidden Services, it's highly unlikely that the researchers were able to capture all of them. While they found that Tor Hidden Services were in fact quite varied in terms of content, the most popular sites in terms of visits were command and control centres for botnets and resources serving adult content. A. Biryukov, I. Pustogarov, and R. Weimann, *Content and Popularity Analysis of Tor Hidden Services.*

137 ***By May 2011, there were***: gawker.com/the-underground-website-where-you-can-buy-any-drug-imag-30818160; www.wired.co.uk/news/archive/2013-10/09/silk-road-guide.

137 ***As altoid suggested, the site***: Vendors were allowed to sell anything they wanted, with a few exceptions. Child pornography, guns and information about other people were banned.

138 ***The site was accessible only***: antilop.cc/sr/files/DPR_Silk_Road_Maryland_indictment.pdf (First indictment).

138 ***In October 2011, altoid returned***: www.thedigitalhq.com/2013/10/03

/silk-road-shut-drugs-hitmen-blunders/: "Who is Silk Road? Some call me SR, SR admin or just Silk Road. But isn't that confusing? I am Silk Road, the market, the person, the enterprise, everything. But Silk Road has matured and I need an identity separate from the site and the enterprise of which I am now only a part. I need a name."

138 ***At that point, a team of between***: The site charged a commission of 10 percent of all sales under $25, which tapers down to 4 percent for anything above $2,500.

138 ***These administrators submitted a "weekly report"***: www.scribd.com/doc/172768269/Ulbricht-Criminal-Complaint.

138 ***Almost 4,000 anonymous vendors had***: www.theguardian.com/technology/2013/nov/25/majority-of-silk-roads-bitcoins-may-remain-unseized; www.theverge.com/2013/10/14/4836994/dont-host-your-virtual-illegal-drug-bazaar-in-iceland-silk-road; www.forbes.com/special-report/2013/silk-road/index.html.

139 ***"We are NOT beasts of burden"***: www.forbes.com/sites/andygreenberg/2013/04/29/collected-quotations-of-the-dread-pirate-roberts-founder-of-the-drug-site-silk-road-and-radical-libertarian/6/; www.forbes.com/sites/andygreenberg/2013/08/14/meet-the-dread-pirate-roberts-the-man-behind-booming-black-market-drug-website-silk-road/.

139 ***Across Tor Hidden Services forums***: One user echoed the experience of many, posting the following in the Silk Road forum: "Like many others here, I discovered and first began using Silk Road because it was a place to get substances I would otherwise not have access to. For a long time that is all it was for me, until I discovered the forums. I truly feel and believe that if communities like this continue to thrive that we will someday change the opinions of those around us the same way my opinion has been altered. Perhaps someday even the 'war on

drugs' will end because the masses will understand us instead of fearing us. To sum up the entirety of this post and to answer your question, the Silk Road, to me, means hope."

140 ***On October 1, 2013*:** edition.cnn.com/2013/10/04/world/americas/silk-road-ross-ulbricht/, see also arstechnica.com/security/2013/10/silk-road-mastermind-unmasked-by-rookie-goofs-complaint-alleges/ and www.bbc.co.uk/news/technology-24371894.

140 ***He had told his housemates*:** www.wired.com/threatlevel/2013/10/ulbricht-delay/. The FBI's investigation was led by Christopher Tarbell, the agent responsible for the 2011 New York sting that caught LulzSec hacker Hector Monsegur (aka Sabu). See the following sources: www.bloomberg.com/news/2013-11-21/silk-road-online-drug-market-suspect-ulbricht-denied-bail-1-; www.slate.com/blogs/crime/2013/11/26/ross_william_ulbricht_redandwhite_did_the_alleged_silk_road_kingpin_lose.html?wpisrc=burger_bar; www.theguardian.com/technology/2013/oct/03/five-stupid-things-dread-pirate-roberts-did-to-get-arrested.

141 ***There swiftly followed the arrest*:** www.telegraph.co.uk/news/uknews/crime/10361974/First-British-Silk-Road-suspects-arrested-by-new-National-Crime-Agency.html; www.theguardian.com/uk-news/2013/oct/08/silk-road-illegal-drugs-arrested-britain; krebsonsecurity.com/2013/10/feds-arrest-alleged-top-silk-road-drug-seller/; www.dailymail.co.uk/news/article-2456758/Two-Dutch-Silk-Road-vendors-alias-XTC-Express-caught-red-handed-layer-MDMA-hair.html?ito=feeds-newsxml.

141 ***Libertas and other site administrators*:** They offered all former Silk Road vendors accounts on the marketplace (vendors have to pay a small bond to be allowed to sell). "We will need to verify that you really were a vendor on S[ilk] R[oad]," wrote Libertas. "To do this, we

ask you PM me with a signed message with your old PGP, linking me to your PGP key on the old forums."

142 ***Inigo, one of Libertas's fellow administrators***: Not everyone was happy that Ulbricht had been arrested. "One day you will be out of jail and I will track you down and demand the $250,000." Many were frustrated that he had seemingly been so lax: "It's almost like he did it for the fame, he wanted to get caught!"; "Running SR and living in the USA?? What the Fuck??"

142 ***"Silk Road has risen"***: twitter.com/DreadPirateSR/status/398117916802961409.

143 ***Buyers and vendors who'd become***: The following is a short timeline of Dark Market drug website activity in the immediate aftermath of the original Silk Road being shut down by the authorities:

October 2, 2013: Silk Road taken down.

October 9: Libertas announces Silk Road 2.0.

October–November: Silk Road's two main rival sites, Black Market Reloaded and the Sheep Market, experience a surge in activity and vendors and buyers shift over.

October: Backopy, the admin of the Black Market Reloaded site, says the site will close after an admin leaked some webpage source code, but subsequently changes his mind when it becomes clear that the source code did not betray any vulnerabilities.

November 6: Silk Road 2.0 goes online. There are new security features, including double validation PGP encryption. It tries to make up for lost ground by validating old vendors automatically.

November 30: Sheep Market shuts down after $5.3 million in Bitcoin is stolen from the site. The site administrators claim that a vendor called EBOOK101 found a bug in the system and stole all of the marketplace's money. Others allege that the administrators absconded with it.

December: The Black Market Reloaded, by now the largest online drugs market, closes. Backopy says they cannot handle the influx of new customers and sellers. Backopy hints at a 2014 relaunch.

December: The administrator of a new site, Project Black Flag, panics and runs off with users' Bitcoins.

December: DarkList, the online drug dealer directory, is launched as a way of keeping track of all the disparate online drugs marketplaces. It closes again in late December.

December: Virginia resident Andrew Michael Jones, Gary Davis from Wicklow, Ireland, and Australian Peter Philip Nash are arrested. The FBI alleges they are the admins of Silk Road 2.0 (Indigo, Libertas and SameSameButDifferent). There is some speculation about FBI infiltration of the site.

December: Agora Market is founded.

January 19, 2014: Drugslist Marketplace starts to offer a new type of security feature called "Multisig escrow."

January 22: DarkList relaunches.

Late January: Cantina Marketplace launches. It is challenged by sceptical Reddit users for security specifics.

Late January (possibly January 27): A group of hackers expose multiple security problems on Drugslist Marketplace. One hacker posts all the site's internal information and user information.

February 2: CannabisRoad is hacked.

February 3: Black Goblin Market launches, and a day later is taken down due to amateurish security.

First week of February: Utopia marketplace is launched. It has a strong connection to Black Market Reloaded.

Early February: The White Rabbit marketplace is set up. It accepts Bitcoins and Litecoins, and runs on I2P, not Tor.

February 12: Dutch police seize Utopia, forcing it offline. They decline to discuss the details.

Early February: Silk Road 2.0 is hacked, $2.7 million in Bitcoins lost.

February 16: Agora Market becomes the most popular marketplace on the deep web.

Late February/early March: Agora is closed down, reopened and closed numerous times as a result of intensive distributed denial of service attacks.

Early March: Hansamarket, a new online drugs market, is opened; it is almost immediately exposed as insecure.

March 19: Pandora Marketplace hacked, $250,000 in Bitcoins are lost. The market stays up.

March 22: EXXTACY Market launched.

March 23: Reddit user "the_avid" exposes EXXTACY as having poor security. The_avid also steals and publishes Red Sun Market server information.

March 24: Serious security issues exposed on White Rabbit Market.

143 ***The* Sydney Morning Herald *warned*:** www.smh.com.au/technology/technology-news/riding-the-silk-road-the-flourishing-online-drug-market-authorities-are-powerless-to-stop-20110830-1jj4d.html (August 30, 2011).

143 ***Charles Schumer, the U.S. senator*:** www.nbcnewyork.com/news/local/123187958.html.

143 ***There are no regulators to turn*:** The greatest scam ever pulled on the original Silk Road was by a vendor called Tony76 who spent months building up a solid online reputation and then ran multiple scams, known as "Finalise Early" scams.

144 ***Major e-commerce companies spend millions*:** N. Nahai, *Webs of Influence*.

145 ***We do this in several different ways*:** allthingsvice.com/2013/04/23/competition-for-black-market-share-hotting-up/.

147 ***It provides the most detailed*:** It was uploaded on a Tor Hidden Service as a very large Excel File. Data relates to all feedback given on the site between January 10, 2014, and April 15, 2014). This data was col-

lected by gathering reviews on all purchases (which is close to mandatory). Not quite back to the level at the height of Silk Road in July 2013—but getting close. One review is left per transaction, not per product, so it is likely that this is a conservative estimate. In 2012, Professor Nicolas Christin wrote an excellent report based on user review feedback left on the original Silk Road; www.andrew.cmu.edu/user/nicolasc/publications/TR-CMU-CyLab-12-018.pdf.

148 ***Although vendors tend to be based***:

Nation ("shipping from")	No. of vendors	Percentage of vendors
United States	231	33
United Kingdom	70	10
Australia	66	9.4
Germany	47	6.7
Canada	36	5.1
Netherlands	36	5.1
Sweden	21	3
Spain	10	1.4
China	9	1.3
Belgium	8	1.2
France	8	1.1
India	8	1.1

148 ***Twenty-one vendors sold over***: Other vendors sell more products—professorhouse is selling 1,170 items—but these are not drugs, and include scamming and hacking guides.

149 ***A very decent salary***:

Name	Products	Overall turnover (99 days)
AmericaOnDrugs	Varied; drugs	$45,209
BlackBazar	Heroin, cocaine, MDMA	$12,068

Koptevo	Prescription drugs only	$9,197
Demoniakteam	Cannabis, ecstasy, psychedelics	$16,287
Instrument	MDMA only	$24,790
California Dreamin	Mainly cannabis, some prescription drugs	$39,329
GucciBUDS	A variety, mainly cannabis	$14,912
MDMAte	MDMA only	$11,727
Aussie Quantomics	Mainly MDMA, some psychedelics	$16,099

149 ***But Silk Road has brought new***: www.redditcom/r/casualiama/comments/1l0axd/im_a_former_silk_road_drug_dealer_ama/; www.vice.com/print/internet-drug-dealers-are-really-nice-guys.

149 ***"We are an importer"***: mashable.com/2013/10/02/silk-road-drug-dealer-interview/.

150 ***"When you're going to leave"***: It is still some way off the complexity of eBay's review system, although eBay has had substantially longer to refine it.

152 ***It didn't matter that no one***: Over time, some vendors built up long-term and sustained reputations on Silk Road. Therefore, many kept the same pseudonym and transferred it to new sites. Libertas—on setting up Silk Road 2.0—allowed all existing Silk Road vendors to immediately become Silk Road 2.0 vendors if their PGP keys matched up. When the marketplace Atlantis went online as a rival to the original Silk Road, verified Silk Road traders were able to become Atlantis traders immediately, in a clever effort to latch on to some of that precious credibility that the Silk Road had facilitated.

153 ***Common tricks include creating fake***: Manipulating the review system is hardly limited to the deep web. The importance of online reviews to

e-commerce is feeding a growing "Online Reputation Management" industry. Hundreds of companies now offer to repair and improve the reputation of companies online. Several major companies have been fined for manipulating or faking their own reviews.

154 ***That's why on the streets*:** M. Daly and S. Sampson, *Narcomania: How Britain Got Hooked on Drugs.*

154 ***Analysis of seized ecstasy tablets*:** Drug Treatment in 2009–10 (Report), National Treatment Agency for Substance Misuse, October 2010; www.unodc.org/documents/data-and-analysis/WDR2012/WDR_2012_web_small.pdf.

154 ***Fourteen died*:** Scottish Drug Forum, "Anthrax and Heroin Users: What Workers Need to Know": www.sdf.org.uk/index.php/download_file/view/262/183/ (accessed April 20, 2014).

154 ***True, price here*:** S. Mahapatra, "Silk Road vs. Street: A Comparison of Drug Prices on the Street and in Different Countries," *International Business Times*, www.ibtimes.com/silk-road-vs-street-comparison-drug-prices-street-different-countries-charts-1414634 (accessed April 20, 2014).

155 ***On the other hand*:** Clarity Way (a drug charity), "The Amazon of Illegal Drugs: The Silk Road vs. The Streets [Infographic]," ClarityWay.com; www.clarityway.com/blog/the-amazon-of-illegal-drugs-the-silk-road-vs-the-streets-infographic/ (accessed April 20, 2014).

155 ***But according to Steve Rolles*:** www.reddit.com/r/Drugs/comments/1tvr4a/the_most_popular_drugs_bought_with_bitcoin_on/cecw84x for some discussion about the quality of the drugs. "With stuff like MDMA in particular, Silk Road was popular because they had the best and the cheapest that most consumers could find. The heroin from the Silk Road was rather expensive compared to local prices, but it was also some of the best (mostly). Couple venders were

selling shit laced with fent[anyl], which just isn't cool, but reviews helped."

157 ***Some of the newer markets***: www.deepdotweb.com/2014/01/25/drugslist-now-offering-full-api-multi-sig-escrow/.

158 ***"[Multi-sig is] the only way"***: www.deepdotweb.com/2014/02/13/silk-road-2-hacked-bitcoins-stolen-unknown-amount/.

158 ***However, researchers have found that***: motherboard.vice.com/blog/bitcoin-isnt-the-criminal-safe-haven-people-think-it-is; anonymity-in-bitcoin.blogspot.com/2011/07/bitcoin-is-not-anonymous.html.

159 ***CoinJoin, for example, works***: bitcointalk.org/index.php?topic=139581.0.

160 ***The future of these markets***: www.chaum.com/articles/Security_Wthout_Identification.htm. This was what David Chaum—the inventor of digital cash twenty years before Satoshi Nakamoto—had in mind all along. In his 1985 book, *Security without Identification: Transaction Systems to Make Big Brother Obsolete*, he set out systems that could combine anonymity with secure payment.

161 ***Dark net markets have introduced***: A. Hirschman, *Exit Voice and Loyalty*.

162 ***When Professor Nicolas Christin analyzed***: www.andrew.cmu.edu/user/nicolasc/publications/TR-CMU-CyLab-12-018.pdf.

162 ***The Silk Road 2.0***: allthingsvice.com/2013/04/23/competition-for-black-market-share-hotting-up/.

163 ***Grams searches the largest markets***: www.deepdotweb.com/2014/04/08/grams-darknetmarkets-search-engine/.

164 ***Illicit substances are more available***: T. Kerr, J. Montaner, B. Nosyk, D. Werb, and E. Wood, "The Temporal Relationship Between Drugs Supply Indicators: An Audit of Internation Government Surveillance Systems," bmjopen.bmj.com/content/3/9/e003077.full.

164 ***Since President Nixon declared war***: "War on illegal drugs failing, medical researchers warn," *BBC News*, October 1, 2013; www.bbc.co .uk/news/uk-24342421.

164 ***There is violence and corruption***: RSA Commission on Illegal Drugs, Communities and Public Policy, "The Supply of Drugs Within the UK," *Drugscope*; www.drugscope.org.uk/Resources/Drugscope/Documents/PDF/Good%20Practice/ supply.pdf.

164 ***The longer the chain***: Peter Reuter, "Systemic Violence in Drug Markets," *Crime, Law and Social Change*, September 2009, vol. 52, issue 3; J. Martin, "Misguided Optimism: the Silk Road closure and the War on Drugs," *The Conversation*, theconversation.com/misguided-optimism-the-silk-road-closure-and-the-war-on-drugs-18937 (accessed April 20, 2014).

164 ***Although reliable figures for***: www.unodc.org/documents/data-and-analysis/WDR2012/WDR_2012_web_small.pdf.

CHAPTER 6: LIGHTS, WEB-CAMERA, ACTION

167 ***The fifteen most visited porn sites***: The following is based on the Alexa Ranking, which ranks websites in terms of their global popularity (a combined score of both page views and unique visitors—in brackets in the list below). They are frequently updated, and were correct as of May 2014. I also searched for the volume of amateur videos available on each of the sites:

Xvideos.com (40): 49,003 videos
Xhamster.com (54): 368,000 videos
Pornhub.com (80): 22,743 videos
Redtube.com (98): 3,517 videos
Xnxx.com (102): 49,011 videos
LiveJasmin.com (107): a cam site, not a porn site; however its popularity is worth noting
Youporn.com (116): 43,597 videos

Tube8.com (213): 49,662 videos
Chaturbate.com (329): another cam site
YouJizz.com (351): 136,883 videos
Motherless.com (359): Motherless is a site dedicated to teen porn; amateur videos make almost all of its content. A search for amateur catalogues 200,413 videos
Beeg.com (362): 3,279 videos
Hardsextube.com (435): 8,450 videos
Drtuber.com (600): 406,119 videos
Nuvid.com (795): 347,112 videos
Spankwire.com (803): 164,111 videos
Sunporno.com (870): 51,397 videos

167 ***The Free Speech Coalition estimated***: Revenue hasn't completely collapsed in the professional porn industry, of course. But the cost of production has tumbled. Back in the early 1980s it cost over $200,000 to produce a standard professional porn film; when video taping turned up in the mid-eighties, the costs fell—but by the mid-nineties a professional porn film usually cost $100,000. Towards the later nineties, some pro-am companies were producing films for closer to $20,000—roughly what it costs now for an average, professional movie. (Although there were still enormous one: *Pirates II*, the most costly porno ever made, had a budget of $8 million.) Figures on porn usage and the size of the industry overall vary a lot. It is a very controversial subject. In 2007, the *Observer* quoted that the industry was worth $13 billion in the United States, the most commonly quoted figure. In 2012, CNBC said the industry was worth $14 billion in the United States. Top Ten Reviews estimates it to currently be worth $57 billion wordwide. In the United States, between 2001 and 2007, internet porn went from a one-billion-dollar industry to a three-billion-dollar industry. See the following: www.theguardian.com/world/2007/dec/16/film.usa; internet-filter-review.toptenreviews.com/internet-pornography-statistics-pg2.html; www.toptenre

views.com/2-6-04.html; www.socialcostsofpornography.com/Doran_Industry_Size_Measurement_Social_Costs.pdf; www.thefreeradical.ca/Toronto_the_naughty.htm; www.overthinkingit.com/2009/03/26/the-adult-film-industry-rediscovers-its-balls/2/.

167 ***In the 1980s, users of***: "Which was to have a private chat wherein you would type in various acts that a 13-year-old boy would want to do with a 13-year-old girl." "BBS life in the 1980s" by Mr Pez, textfiles.com/history/golnar.txt.

167 ***The first erotic stories Usenet group***: www.asstr.org/~apuleius/assh-faq.html#2.

168 ***Porn is now estimated***: R. McAnulty and M. Burette, *Sex and Sexuality*, vol. 1; articles.orlandosentinel.com/1998-03-28/lifestyle/9803270925_1_entertainment-online-video, p. 269; www.bbc.co.uk/news/technology-23030090.

168 ***Jennifer was the very first***: T. M. Senft, *Camgirls: Celebrity and Community in the Age of Social Networks*, p. 44.

168 ***At its peak, four million***: "Behind the Scenes with Jennifer Ringley," promo for *Web Junk Presents . . . 40 Greatest Internet Superstars*, March 18, 2007; www.spike.com/video-clips/po0d6t/behind-the-scenes-with-jennifer-ringley (accessed December 4, 2013).

168 ***In 1998, she divided her site***: L. Green, *The Internet: An Introduction to New Media*; www.yorku.ca/robb/docs/camgi.pdf.

168 ***They included several spoof JenniCam***: www.naturistplace.com/wnl-0101.htm.

169 ***Camming was becoming a gainful***: www.nytimes.com/2005/12/19/national/19kids.ready.html?pagewanted=7&_r=0&ei=5090&en=aea51b3919b2361a&ex=1292648400&partner=rssuserland&emc=rss.

170 ***Today there are probably around***: The number 50,000 is the closest I've got to an accurate estimate. It was provided by the administrator

of a camgirl support community. There are also a certain number of camgirls from Eastern Europe and South America, and from South East Asia. Chaturbate has recently stopped accepting girls from the Philippines, for fear that they are being exploited.

170 ***According to* The New York Times:** www.nytimes.com/2013/09/22/technology/intimacy-on-the-web-with-a-crowd.html?adxnnl=1&pagewanted=all&adxnnlx=1394884188-8+B9Okpt1TokwE/tHhXoAw; www.theverge.com/2013/9/23/4761246/cam-sex-is-booming-business-for-porn-industry.

170 ***There is even a large*:** One of the support groups is called WeCamGirls. It started up in 2012. That year, the site had around 100,000 visitors. In 2013, it was more than twice that. It currently has over 3,000 active members.

171 ***She then joined Chaturbate*:** www.wecamgirls.com/articles/in-the-spotlight-with-cliche/.

171 ***"I guess I'm a real person"*:** "Labours of Love: Netporn, Web 2.0 and the Meanings of Amateurism," *New Media & Society*, vol. 12, no. 8, 2010.

171 ***"It's a better kind of porn"*:** F. Attwood, *Porn.com*, p. 139, nms.sagepub.com/content/early/2010/06/08/1461444810362853.full.pdf.

171 ***Around the world there are*:** www.emarketer.com/Article/Where-World-Hottest-Social-Networking-Countries/1008870.

171 ***Some psychologists think that social*:** M. Ma, "Understanding the Psychology of Twitter," *Psychology Today*, March 27, 2009; www.psychologytoday.com/blog/the-tao-innovation/200903/understanding-the-psychology-twitter (accessed December 5, 2013); journal.frontiersin.org/Journal/10.3389/fnhum.2013.00439/full#h2.

172 ***Sharing our every intimacy is*:** www.psychologytoday.com/blog/the-tao-innovation/200903/understanding-the-psychology-twitter;

Panek et al., "Mirror or Megaphone? How Relationships Between Narcissism and Social Networking Site Use Differ on Facebook and Twitter," *Computers in Human Behaviour*, vol. 29; www.sarakonrath.com/media/publications/narcissism_SNS_sites.pdf; www.jcr-admin.org/files/pressPDFs/112712180022_Stephen_Article.pdf.

172 ***She records young people spending***: S. Turkle, *Alone Together*, p. 180.

172 ***Pew Research's 2013 study***: "Prevalence and Characteristics of Youth Sexting: A National Study," September 19, 2011, www.pewinternet.org/2014/02/11/main-report-30/; "Basically . . . porn is everywhere," p. 28.

173 ***That's not to say young***: D. Boyd, *It's Complicated*, p. 57.

173 ***Sharing images of ourselves***: www.wired.com/wiredscience/2012/05/opinion-naked-sexting/.

173 ***Viewers respond—sometimes positively***: Like most online communities, there are rules: "no random porn dumps," "post pictures of *yourself!*" and of course, "be respectful to each other."

175 ***In 2011, a Facebook group***: Facebook was originally called Facemash. Mark Zuckerberg and his university friends wanted to rate the pictures of female students they'd managed to grab—without permission, of course—from the Harvard University files. Facemash placed a photo of female students next to each other and asked users to vote on who they thought was the best looking, with an algorithm slowly pushing certain girls up or down the list. "One thing is certain," wrote Zuckerberg on his personal blog at the time, "and that's that I'm a jerk for making this site. Oh well, someone had to do it eventually." He was right. Within four hours, there had been 22,000 visits to his site.

175 ***At the moment of writing***: abcnews.go.com/blogs/headlines/2012/03/facebook-shuts-down-most-beautiful-teen-page/; toronto.ctvnews.ca/cutest-teens-2013-facebook-page-taken-down-1.1540454; www.facebook.com/CutestTeensOfficialPage?fref=ts.

176 ***We are already thirty minutes late***: Vex wears gray knee-length socks, gray knickers, and a gray top with her midriff showing. So does Blath. Auryn wears something similar but in black.

184 ***But there are other girls***: www.reddit.com/r/IAmAcomments/1hq5t9/.

184 ***"Have you heard of this"***: While I was writing this chapter, Chaturbate started accepting Bitcoin.

188 ***The growing volume of sexually***: D. K. Citron and M. Franks, "Criminalising Revenge Porn," digitalcommons.law.umaryland.edu/cgi/view content.cgi?article=2424&context=fac_pubs.

188 ***It was found that he had***: oag.ca.gov/news/press-releases/attorney-general-kamala-d-harris-announces-arrest-revenge-porn-website-operator.

188 ***Similar things are happening in***: www.dailymail.co.uk/femail/article-2175591/Is-YOUR-child-sending-sex-texts-school.html#ixzz26L5qLbJh; www.nytimes.com/2011/03/27/us/27sexting.html?pagewanted=all&_r=0. The statistics are cited in Citron and Franks, "Criminalising Revenge Porn."

189 ***Jessica Logan from Ohio committed***: K. Albin, "Bullies in a Wired World: The Impact of Cyberspace Victimization on Adolescent Mental Health and the Need for Cyberbullying Legislation in Ohio," *Journal of Law and Health*, vol. 25, iss. 1, pp. 155–90.

189 ***In another American high school***: "3 Juveniles Accused of Sexually Exploiting Female Classmates"; www.newschannel5.com/story/21890716/3-juveniles-accused-of-sexually-exploiting-female-classmates.

190 ***In his famous 1990 article***: Howard Rheingold, "Teledildonics: Reach Out and Touch Someone"; janefader.com/teledildonics-by-howard-rheingold-mondo-2000-1990/.

191 ***"We're not a community"***: twitlonger.com/show/n_1s0rnva, by @thecultofleo.

CHAPTER 7: THE WERTHER EFFECT

194 ***Eighteen percent of U.S.***: www.pewinternet.org/2011/02/28/peer-to-peer-health-care-2/.

194 ***Studies consistently find that speaking***: www.mind.org.uk/media/418956/Peer-Support-Executive-Summary-Peerfest-2013.pdf, p. 2; www.mentalhealth.org.uk/help-information/mental-health-a-z/P/peer-support/. See also Mental Health Foundation, *The Lonely Society?*

194 ***Arguably the first "pro" self-harm***: ash.notearthday.org/charter.html; a.s.h. is sometimes called a website, but it is in fact a Usenet *alt.** group hierarchy group, which has no central authority, meaning it is difficult to close down.

194 ***The first two posts on a.s.h.***: Incidentally, this is one of the great myths about suicides. The rate of suicide in the United States and most of the Northern Hemisphere is lower in December than any other month. There are several excellent academic studies that debunk this myth. The two halves of this quote were technically separate posts, but are considered to be part of the "founding charter" of a.s.h.

195 ***Today there are hundreds of***: Beals would later explain that a.s.h. became something that he'd never anticipated. www.zenithcitynews.com/010411/feature.htm.

195 ***The number has steadily increased***: "Self-Injury and the Internet: Reconfiguring the Therapeutic Community," in *RESET*, vol. 1, no. 2, 2013.

195 ***A 2007 study examining***: E. Bond, *Virtually Anorexic—Where's the Harm?*, www.ucs.ac.uk/Faculties-and-Centres/Faculty-of-Arts-Business-and-Applied-Social-Science/Department-of-Children-Young-People-and-Education/Virtually%20Anorexic.pdf; bir.brandeis.edu/bitstream/handle/10192/24532/BeliveauThesis2013.pdf?sequence=1.

198 ***I found hundreds of Tumblr***: And although many of these platforms ban content that glorifies eating disorders and self-harm, it is a rule that is extremely difficult to enforce. When Instagram banned users from searching for the hashtag "thinspo," users started writing it as "th1nspo" instead.

205 ***But providing information or discussing***: www.ealaw.co.uk/media/uploaded_files/circular-03-2010-assisting-encouraging-suicide.pdf; see "Encouraging or Assisting Suicide: Implementation of Section 59 of the coroners and Justice Act 2009," *MoJ Circular 2010/03*, January 28, 2010. The law applies the same online as off. "Subsection (2) of section 59 sets out the single offence which replaces the offences of aiding, abetting, counselling or procuring suicide and of attempting to do so. The offence will apply where a person does an act which is capable of encouraging or assisting another person to commit or attempt to commit suicide, and intends his act to so encourage or assist. But simply providing information about or discussing the issue of suicide through the internet (or any other medium), where there is no such intent, is not an offence." www.cps.gov.uk/publications/prosecution/assisted_suicide_policy.html; lostallhope.com/suicide-statistics; www.cdc.gov/violenceprevention/pdf/Suicide_DataSheet-a.pdf.

206 ***More than two thirds were***: E. Bale, H. Bergen, D. Casey, K. Hawton, A. Shepherd, and S. Simpkin, "Deliberate Self-Harm in Oxford 2008," Centre for Suicide Research.

208 ***Within hours, Amelia and dozens***: This specific case was discovered and set out by Emma Bond in her book, *Virtually Anorexic*.

209 ***This strange phenomenon became known***: informahealthcare.com/doi/pdf/10.1080/08039480410005602.

210 ***The month after the August***: L. Coleman, *The Copycat Effect: How the*

Media and Popular Culture Trigger the Mayhem in Tomorrow's Headlines, Paraview and Pocket Books, p. 2.

210 ***In the 1980s, a number***: www.psychologytoday.com/blog/media-spotlight/201208/when-suicides-come-in-clusters; railwaysuicideprevention.com/scientific-literature-review/choose.html; ccp.sagepub.com/content/12/4/583.full.pdf+html; www.theguardian.com/lifeandstyle/2009/mar/01/bridgend-wales-youth-suicide-media-ethics; www.people.com/people/archive/article/0,,20595753,00.html.

210 ***The same phenomenon has been***: www.ivonnevandevenstichting.nl/docs/SuicideAndTheMedia.pdf; N. Kristakis, "Connected"; content.time.com/time/health/article/0,8599,1808446,00.html. It's also true in less damaging circumstances. If those around us are overweight, we're more likely to pile on the pounds. Canned laughter on television makes us more likely to laugh. If we're told how many other hotel guests reuse their towels, we're more likely to do the same, www.media-studies.ca/articles/influence_ch4.htm; www.lse.ac.uk/GranthamInstitute/publications/WorkingPapers/Papers/130-39/WP133-Exploring-beliefs-about-bottled-water-intentions-to-reduce-consumption.pdf; www.otismaxwell.com/blog/persuading-people-social-proof/.

210 ***As a result, many countries***: The Samaritans released an excellent report of guidelines in 2013, specifically designed to reduce copycat suicides. The Press Complaints Commission Released a special briefing note on press coverage of suicide in March 2009, www.pcc.org.uk/advice/editorials-detail.html?article=NTU4MQ==.

211 ***"Have a nice life!"***: archive.ashspace.org/ash.xanthia.com/conibear.html.

212 ***Academic research has found that***: J. Taylor et al., "Motivations for Self-Injury, Affect, and Impulsivity: A Comparison of Individuals

with Current Self-Injury to Individuals with a History of Self-Injury," *Suicide and Life-Threatening Behaviour*, vol. 42, no. 6, December; K. Hawton et al., "Self-harm and Suicide in Adolescents," *The Lancet*, vol. 379, iss. 9834, is another excellent source, onlinelibrary.wiley.com/doi/10.1111/j.1943-278X.2012.00115.x/pdf.

213 ***As he recovered in the hospital***: The footage is still, at the time of writing, available on the website live leak. The four chat quotes were collected by the author freeze-framing the video footage each time the filmer returned to update the /b/ board. news.nationalpost.com/2013/12/04/university-of-guelph-student-who-attempted-to-take-his-life-on-internet-video-now-cyberbullied-on-facebook/.

213 ***Suicide, self-harming, and eating disorders***: There are serious difficulties in recording incidences of self-harm, and it is widely considered to be under-reported. A similar thing occurs in respect of suicide, which coroner's reports will often not include. There is a considerable literature available regarding the UK, however. See, for example, the following: D. Gunnell et al., "The Use of the Internet by People who Die by Suicide in England: A Cross Sectional Study," *Journal of Affective Disorders*, vol. 141, pp. 480–83. See also: www.rcpsych.ac.uk/files/pdfversion/cr158.pdf; www.telegraph.co.uk/women/womens-health/4682209/Anorexic-girls-admitted-to-hospital-rise-by-80-per-cent-in-a-decade.html; www.publications.parliament.uk/pa/cm200607/cmhansrd/cm070606/text/70606w0016.htm. A 2013 King's College study in the *British Medical Journal* concluded that between 2000 and 2009, anorexia and bulimia rates remained stable. However, it also noted that British definitions of these eating disorders are different to those in the United States. There were between 1 and 2 instances of bulimia per 100,000 for males, and between 21 and 25 per 100,000 for females between 2000 and 2009

in the UK. There were between 0.2 and 2 instances of anorexia in males per 100,000, and between 11 and 17 cases of anorexia in females per 100,000 between 2000 and 2009 in the UK. However, diagnoses of "eating disorder not otherwise specified" increased by 60 percent in females, from just under 18 per 100,000 in 2000 to 28.4 per 100,000 in 2009, and by 24 percent in males, from 3.4 to 4.2 per 100,000: bmjopen.bmj.com/content/3/5/e002646.full.?rss=1#F1; telegraph.co.uk/graphics/projects/inside-the-world-of-anorexia-blogging/; www.telegraph.co.uk/health/healthnews/10607237/Eating-disorder-increase-among-young-people.html; www.kcl.ac.uk/iop/news/records/2013/May/Eating-disorders-increase.aspx.

214 ***Sites and forums that reduce***: Mental Health Foundation, *The Lonely Society?*

214 ***Being a nurse, Cami could***: This is taken from relevant court documents: www.mncourts.gov/Documents/0/Public/Clerks_Office/Opinions/opa110987-071712.pdf and www.margaretdore.com/pdf/melchert-dinkel_ff_etc_001.pdf.

217 ***Cami was a middle-aged man***: N. Labi, "Are You Sure You Want to Quit the World?," *GQ*, October 2010, www.gq.com/news-politics/newsmakers/201010/suicide-nurse-mark-drybrough-chatrooms-lidao (accessed December 5, 2013).

CONCLUSION: ZOLTAN VS. ZERZAN

219 ***In Plato's* Phaedrus, *Socrates worried***: Plato, *Phaedrus*, translated by B. Jowett (Project Gutenberg, October 30, 2008).

219 ***When books began to roll off***: V. Bell, "Don't Touch that Dial! A History of Media Technology Scares, from the Printing Press to Facebook," *Slate*, February 15, 2010, www.slate.com/id/2244198.

219 ***Although Marconi believed his radio***: "Radio in the 1920s: Collected commentary," National Humanities Center, americainclass.org

/sources/becomingmodern/machine/text5/colcommentaryradio.pdf. Also T. O'Shei, *Marconi and Tesla: Pioneers of Radio Communication.*

220 ***It would, he believed*:** J. R. Pierce, "Communication," *Scientific American*, vol. 227, no. 3, September 1972, cited in M. Hauben and R. Hauben, *Netizens: On the History and Impact of Usenet and the Internet*, p. 56. M. Greenberger (ed.), *Management and the Computer of the Future*, www.kurzweilai.net/memorandum-for-members-and-affiliates-of-the-intergalactic-computer-network; J.C.R. Licklider quoted in B. Woolley, *Virtual Worlds: A Journey in Hype and Hyperreality*, www.livinginternet.com/i/ii_licklider.htm; Licklider also quickly appreciated the power of computer networks, and predicted the effects of technological distribution, describing how the spread of computers, programs and information among a large number of computers connected by a network would create a system more powerful than could be built by any one organization. In August 1962, Licklider and Welden Clark elaborated on these ideas in the paper "On-Line Man Computer Communication," one of the first conceptions of the future internet.

220 ***Anarchists dreamed of a world*:** Cited in M. Dery, *Escape Velocity*, p. 45.

220 ***The early nineties were ablaze*:** There is an excellent resource which documents 1990s predictions about the future of the internet, here: www.elon.edu/predictions/early90s.

220 ***"We're talking about Total Possibilities"*:** *Mondo 2000*, no. 1, 1989.

221 ***Nicholas Negroponte—former director*:** motspluriels.arts.uwa.edu.au/MP1801ak.html.

221 ***"They gaze on technology as"*:** www.nyu.edu/projects/nissenbaum/papers/The-Next-Digital-Decade-Essays-on-the-Future-of-the-Internet.pdf.

221 ***Others were concerned that we*:** T. Furness and J. Lanier, *Are New*

Realities More or Less Real?; M. Heim, *Scholars Try to Measure the Impact*; G. Celente, *Online, All the Time: Today's Technology Makes the Office Omnipresent, but Is That Any Way to Live?*

223 ***"By thoughtfully, carefully and yet"***: M. More, "The Philosophy of Transhumanism," in M. More and N. Vita-More, *The Transhumanist Reader: Classical and Contemporary Essays on the Science, Technology and Philosophy of the Human Future*, p. 4.

223 ***Nick Bostrom, a well-known***: www.nickbostrom.com/papers/history.pdf.

223 ***In 1993, Vernor Vinge popularized***: "The Coming Technological Singularity: How to Survive in the Post-Human Era," available here: www.rohan.sdsu.edu/faculty/vinge/misc/singularity.html; I. J. Good, "Speculations Concerning the First Ultraintelligent Machine," *Advances in Computers*, vol. 6.

224 ***By 1998, the burgeoning group***: www.fhi.ox.ac.uk/a-history-of-transhumanist-thought.pdf; M. More and N. Vita-More, *The Transhumanist Reader*, pp. 54–55. Even before this was the lesser known "Transhuman Manifesto," penned in 1983 by Natasha Vita-More.

229 ***In 2012, in Essen***: www.wired.co.uk/news/archive/2012-09/04/diy-biohacking.

233 ***A growing number of writers***: For "Technostress," see W. Powers, *Hamlet's Blackberry*. The other terms are respectively attributable to William van Winkle, David Lewis, Eric Schmidt, and Leslie Perlow. See J. Schumpeter, "Too Much Information: How to Cope with Data Overload," *Economist*, June 30, 2011. www.rcpsych.ac.uk/files/pdfversion/cr158.pdf.

FURTHER READING

For anyone wishing to explore the subjects and issues raised in *The Dark Net* in greater detail, there are many excellent books, articles and web resources available on each of the themes I covered. The following list will offer, I hope, a useful starting point.

INTRODUCTION: LIBERTY OR DEATH

Bell, J. *Assassination Politics*, www.outpost-of-freedom.com/jimbellap.htm.

Boyd, D. *It's Complicated: The Social Lives of Networked Teens*. An incredibly useful and clear-eyed account of young people's relationship with social networks.

Hafner, K., and M. Lyon. *When Wizards Stay Up Late: The Origins of the Internet*.

Krotoski, A. *Untangling the Web: What the Internet Is Doing to You*.

Pariser, E. *The Filter Bubble: What the Internet Is Hiding from You*.

Suler, J. "The Online Disinhibition Effect," in *CyberPsychology and Behaviour*. An extremely influential theory about what effect that communicating from behind a screen has on us.

Turkle, S. *The Second Self: Life on the Screen and Alone Together*. Sherry

Turkle is without question one of the world's experts on this subject, and someone whose studies on the impact of computers on human behavior and identity are required reading.

Zuckerman, E. *Rewire: Digital Cosmopolitans in the Age of Connection.*

CHAPTER 1: UNMASKING THE TROLLS

Coleman, G. *Our Weirdness Is Free.*

Olson, P. *We Are Anonymous.* This and the above are both excellent accounts of the hacktivist group Anonymous, of the evo-lution of /b/, and of trolling more generally.

Phillips, W. "LOLing at Tragedy: Facebook, Memorial Trolls and Reaction to Grief Online," *First Monday* vol. 16, no. 12, firstmonday.org/ojs /index.php/fm/article/view/3168. Whit-ney Phillips is one of the few academics who specializes in trolling.

Schwartz, M. "The Trolls Among Us," *New York Times*, August 3, 2008. One of best articles about trolling and trolling sub-cultures, featuring the notorious hacker and troll "weev."

The websites Encyclopedia Dramatica and KnowYourMeme are both excellent resources for trolling culture: but enter these sites at your own risk. Examples of trolling on Bulletin Board Systems are usefully archived on the website textfiles.com; while there are several online resources dedicated to Usenet trolling groups, notably those maintained by Ken Hollis at www.digital.net/~gandalf/.

CHAPTER 2: THE LONE WOLF

Bartlett, J., and M. Littler. *Inside the EDL*, Demos.

Bergen, J., and W. Strathern. *Who Matters Online: Measuring Influence, Evaluating Content and Countering Violent Extremism in Online Social Networks,* International Centre for the Study of Radicalisation.

Conway, M., et al. "Uncovering the Wider Structure of Extreme Right Communities Spanning Popular Online Networks."

Copsey, N. *The English Defence League*, Faith Matters, faith-matters.org /images/stories/fm-reports/english-defense-league-report.pdf.

Simon, J. *Lone Wolf Terrorism: Understanding the Growing Threat.*

CHAPTER 3: INTO GALT'S GULCH

Greenberg, A. *This Machine Kills Secrets: Julian Assange, the Cypherpunks, and Their Fight to Empower Whistleblowers.* An invaluable guide to cypherpunk technology and ideology, and the significance of the cypherpunk philosophy to whistleblowers.

Levy, S. "Crypto-rebels" in *Wired* and *Crypto: How the Code Rebels Beat the Government Saving Privacy in a Digital Age.* "Crypto-rebels" was the first mainstream account of the cypherpunks; while *Crypto* remains the best account of the movement overall.

Manne, R. "The Cypherpunk Revolutionary: Julian Assange," in *Making Trouble: Essays Against the New Australian Complacency.*

May, T. *Cyphernomicom*, Tim May's book-length essay, providing an excellent insight into the cypherpunk philosophy. Available here: www.cypherpunks.to/faq/cyphernomicron/cyphernomicon.html. For the technically minded, David Chaum's paper "Security Without Identification: Transaction Systems to Make Big Brother Obsolete" is perhaps the most important single document in terms of understanding the mathematics of the cypherpunk movement. Most of the Cypherpunk and Metzdowd Cryptography Mailing List posts are archived and available online: cypherpunks.venona.com/ and www.metzdowd.com/pipermail/cryptography/.

CHAPTER 4: THREE CLICKS

Davidson, J., and P. Gottschalk. *Internet Child Abuse: Current Research and Policy.*

Finkelor, D. *A Sourcebook on Child Sexual Abuse.* A work which sets out the "classic" model of child grooming, written before social networks were widely used.

Martellozzo, E. "Children as Victims of the Internet: Exploring Online Child Sexual Exploitation."

Ogas, O., and S. Gaddam. *A Billion Wicked Thoughts.* An unusual and extremely valuable piece of detailed work into people's sexual desires based on internet search terms.

Wortley, R., and S. Smallbone. *Internet Child Pornography: Causes, Investigation and Prevention.* An excellent overview of how the advent of the internet radically transformed child pornography. Following the ups and downs on the Tor Hidden Service child pornography is extremely difficult, and has not been written about in detail. *The Daily Dot*, *Gawker*, *Wired*, and *Vice* all have extremely useful articles and reports on the subject. In particular: Patrick Howell O'Neill's articles on Freedom Hosting: www.dailydot.com/news/eric-marques-tor-freedom-hosting-child-porn-arrest/; Adrian Chen's articles on "Operation DarkNet": gawker.com/5855604/elaborate-anonymous-sting-snags-190-kiddie-porn-fans; and *Wired*'s series "Threat Level": www.wired.com/category/threatlevel/.

CHAPTER 5: ON THE ROAD

Christin, N. "Travelling the Silk Road: A Measurement Analysis of a Large Anonymous Online Market Place." An innovative academic study of the volume of trading on the Silk Road in 2012.

Daly, M., and S. Sampson. *Narcomania: How Britain Got Hooked on Drugs.*

European Monitoring Centre for Drugs and Drug Addiction, *European Drugs Report 2013: Trends and Developments.*

Powers, M. *Drugs 2.0.* An excellent account of how the internet has changed the way drugs are bought and sold.

Criminal Complaint documents are extremely valuable sources, including those laid against Ulbricht himself. As usual, online magazine tend to have the most detailed accounts of the Silk Road and the other dark net markets. Of particular note are Adrian Chen's 2011 *Gawker* article "The Underground Web-site Where You Can Buy Any Drug Imaginable," gawker.com/the-underground-website-where-you-can-buy-any-drug-imag-30818160 (which was the first significant journalistic exploration on the subject), and Eileen Ormsby's excellent blog AllThingsVice, which covers several aspects of the deep web, but especially the dark net markets, allthingsvice.com/about/.

CHAPTER 6: LIGHTS, WEB-CAMERA, ACTION

Attwood, F. *Porn.com: Making Sense of Online Pornography.* An excellent overview of how the internet is changing porn-ography, and in particular how home-made pornography changes the relationship between viewer and producer.

Boellstorff, T. *The Coming of Age in Second Life*. One of the first detailed anthropological studies of life in virtual worlds, and indispensable for understanding how virtual-world avatars reflect their user's behaviors.

Panek, E., et al. "Mirror or Megaphone? How Relationships between Narcissism and Social Networking Site Use Differ on Facebook and Twitter," *Computers in Human Behaviour*, vol. 29.

Senft, T. *Camgirls: Celebrity and Community in the Age of Social Networks*. The first detailed, and still most comprehensive, study of cam-models.

CHAPTER 7: THE WERTHER EFFECT

Adler, P., and P. Adler. *The Tender Cut: Inside the Hidden World of Self-Injury*. An excellent and very detailed study of self-injury, in particular cutting.

Barak, A. "Emotional Support and Suicide Prevention through the Internet: A Field Project Report," *Computers in Human Behaviour*, vol. 23, and "Suicide Prevention by Online Support Groups: An Action Theory–Based Model of Emotional First Aid," *Archives of Suicide Research*, vol. 13, no. 1 (2009). Barak's work provides a critical analysis of the effect of online suicide support groups on users.

Bond, E. *Virtually Anorexic—Where's the Harm?* An accessible but rigorous study of the scale, type, and content of pro-ana sites, and an excellent introduction to the subject.

Coleman, L. *The Copycat Effect: How the Media and Popular Culture Trigger the Mayhem in Tomorrow's Headlines*.

Gunnell, D., et al. "The Use of the Internet by People who Die by Suicide in England: A Cross-Sectional Study," *Journal of Affective Disorders*, vol. 141.

Hawton, K., et al. "Self-harm and Suicide in Adolescents," *The Lancet*, vol. 379, iss. 9834.

Mental Health Foundation, *The Lonely Society*.

Montgomery, P., et al. "The Power of the Web: A Systematic Review of Studies of the Influence of the Internet on Self-Harm and Suicide in Young People," *PLoS ONE*.

Sueki, H. "The Effect of Suicide-Related Internet Use on Users' Mental Health: A Longitudinal Study," *Journal of Crisis Intervention and Suicide Prevention*, vol. 34(5).

CONCLUSION: ZOLTAN VS. ZERZAN

Istvan, Z. *The Tranhumanist Wager*. Zoltan's loosely autobiographical work, which sets out a picture of a fairly bleak near future in which transhumanists go to war with the rest of the world.

More, M., and N. Vita-More (eds.). *Transhumanist Reader*. An excellent overview of some of the more technical and philosophical aspects of the transhumanist movement, edited by two leading exponents. It includes a chapter by Anders Sandberg on mind uploading.

Naughton, J. *From Gutenberg to Zuckerberg: What You Really Need to Know About the Internet*.

Segal, H. *Technological Utopianism in American Culture*.

Zerzan, J. *Future Primitive and Future Primitive Revisited*. A handy introduction into the anarcho-primitivist philosophy, and Zerzan's most well-known written work.

ABOUT THE AUTHOR

Jamie Bartlett is the Director of the Centre for the Analysis of Social Media at the think tank Demos, where he specializes in online social movements and the impact of technology on society. He writes about technology for *The Telegraph* and is a frequent commentator for media outlets throughout the world. He lives in London.